UP-TO-SPEED

# OPERATIONS

by Andrew M. Schorr
and Thomas H. Hatch

# UP-TO-SPEED MATH

**Number Sense**

**Operations**

**Working with Fractions, Decimals, and Percents**

**Geometry and Measurement**

**Patterns, Functions, and Algebra**

**Data Analysis and Reasoning**

created by **Kent Publishing Services, Inc.**
designed by **Signature Design Group, Inc.**

SADDLEBACK PUBLISHING, INC.
Three Watson
Irvine, CA 92618-2767

E-mail: info@sdlback.com
Website: www.sdlback.com

ISBN 1-56254-363-6

Printed in the United States of America.

1 2 3 4 5 6 PP 08 07 06 05 04 03

# TABLE OF CONTENTS

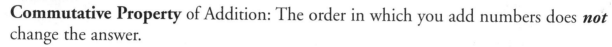

# BASIC PROPERTIES OF ADDITION

**Commutative Property** of Addition: The order in which you add numbers does *not* change the answer.

4 + 3 = 7          27 + 32 = 59

3 + 4 = 7          32 + 27 = 59

**Associative Property** of Addition: The way you group numbers to add them does *not* change the answer.

6 + 9 + 5 + 2 = 22          29 + (43 + 68) = 140

(6 + 9) + (5 + 2) = 22          (29 + 43) + 68 = 140

6 + (9 + 5) + 2 = 22

**Zero Property** of Addition: Adding zero to a number does *not* change the answer.

2 + 0 = 2

17 + 0 = 17

157 + 0 = 157

 **A**     Use the chart below to respond to the items that follow. Show your work.

1. Add the number of wins for the Marlins and Braves.

   Reverse the order of the numbers, and add again.

   What property does this illustrate?

   _____

| Baseball Team Wins | |
|---|---|
| **Team** | Wins |
| **St. Louis Cardinals** | 50 |
| **Cleveland Indians** | 43 |
| **Florida Marlins** | 42 |
| **Minnesota Twins** | 37 |
| **Anaheim Angels** | 43 |
| **San Francisco Giants** | 42 |
| **Detroit Tigers** | 37 |
| **Atlanta Braves** | 50 |

2. Write a number sentence to show that the Cardinals' number of wins will remain the same if the team does not win the next game.

_____

What property does this illustrate? _____

3. Write two number sentences: (1) Add the wins of the Indians and Twins, then add that total to the Giants' wins. (2) Add the wins of the Giants and Twins, then add that total to the Indians' wins.

(1) _____   (2) _____

What property does this illustrate? _____

**B** Complete the number sentences.

1. 3,576 + 7,542 = 7,542 + _____

2. 476 + 0 = _____

3. 1,567 + (4,555 + 879) = (1,567 + 4,555) + _____

4. (8,925 + 543) + 1,115 = 8,925 + (543 + _____)

**C** What property does each number sentence illustrate?

1. 4,123 = 0 + 4,123 _____

2. 964 + 321 = 321 + 964 _____

3. (6,000 + 2,000) + 4,000 = 6,000 + (2,000 + 4,000) _____

4. 8,154 + 0 = 8,154 _____

**D** Write a number sentence based on the property given in parentheses. The first one is done for you.

1. 45, 18, 3 (commutative) _45 + 18 + 3 = 3 + 18 + 45_____

2. 68, 143, 96 (associative) _____

3. 1,654, 544, 3,675 (associative) _____

4. 78, 0 (zero) _____

5. 651, 492, 567 (commutative) _____

6. 0, 830 (zero) _____

# ADDING TWO NUMBERS

**Problem Solving Strategy**
**Solve word problems by writing a number sentence first.**
**Use *only* the numbers you need to solve the problem.**

Wyoming's capital is Cheyenne. Wyoming is the state with the fewest number of people, 479,602. Wyoming became a state in 1890. California's capital is Sacramento. California became a state in 1850 and has the greatest population with 33,145,121. What is the total population of the two states?

Wyoming    California    total

479,602 + 33,145,121 = <u>33,624,723</u>

**To add numbers, line them up starting at the one's place.**

| ten millions | millions | hundred thousands | ten thousands | thousands | hundreds | tens | ones |
|---|---|---|---|---|---|---|---|
|  |  | 4 | 7 | 9 | 6 | 0 | 2 |
| +3 | 3 | 1 | 4 | 5 | 1 | 2 | 1 |
| 3 | 3 | 6 | 2 | 4 | 7 | 2 | 3 |

 Use the chart below to answer the questions that follow. Show your work.

1. Missouri has **4,585,559** more people than Montana. What is the population of Missouri?

| Estimated Population 1999 ||
|---|---|
| **State** | Population |
| **Oregon** | 3,316,154 |
| **Delaware** | 753,538 |
| **Montana** | 882,779 |
| **Michigan** | 9,863,775 |
| **New Mexico** | 1,739,844 |

*Source: U.S. Census Bureau*

2. What would the total population of Michigan be if it increases by 29,278 people?

3. If the population of New Mexico increases by 300,000, the state will receive additional highway funds from the federal government. What would the new population total be?

**B** Find the sums.

1. 2,987,341 + 39,065 = _____

2. 164,894 + 1,765,644 = _____

3. 4,755,923 + 1,977 = _____

4. 6,466,679 + 2,563,957 = _____

5. 678,487 increased by 392,478 is _____

6. 3,563,788 increased by 8,999 is _____

7. 16,477 increased by 876,471 is _____

8. 6,813,789 increased by 2,234,865 is _____

**C** Solve. Show your work.

1. Alaska and Hawaii were the last two states admitted into the United States of America. Alaska covers 586,400 square miles and has a population of 619,500. Hawaii covers 6,450 square miles and has a population of 1,185,497. What is the total population of the two states?

2. In 1998, the population of Connecticut was 3,272,563. By 1999, the population had increased by 9,468. What was Connecticut's population in 1999?

# ADDING MORE THAN TWO NUMBERS

**Problem Solving Strategy**
**Use estimation to check the reasonableness of your answers when adding more than two numbers.**

Loretta records the number of cans students collect for school recycling. The first week they collect 5,678 cans. The second week they collect 3,943 cans. The third week they collect 2,389 cans. How many cans do they collect in three weeks?

**Round each number to the nearest 1,000 to check the reasonableness of your answer.**

| Actual Sum | Estimated Sum |
|:---:|:---:|
| 5,678 | 6,000 |
| 3,943 | 4,000 |
| 2,389 | 2,000 |
| 12,010 | 12,000 |

 **A** Answer the questions using the chart below. Estimate to check the reasonableness of your sums.

Many people go to the movies. The chart at the right shows the box office sales for some of the most popular movies of recent years.

| Movie Title | Box Office Sales |
|:---:|:---:|
| Jurassic Park | $920,100,000 |
| Independence Day | $811,200,000 |
| Lion King | $767,900,000 |
| Home Alone | $533,800,000 |
| Toy Story 2 | $485,600,000 |

*Source: http://www.worldwideboxoffice.com*

1. *Lion King, Home Alone,* and *Toy Story 2* are favorite movies for many students.

   How much did they earn altogether? _____

2. What were the total sales for the five movies on the chart?

   _____

**B** Add the numbers and write the correct sign <, >, or = in the circles to complete the number sentences.  Show your work.

1. $6,433 + 5,976 + 5,398 \bigcirc 4,654 + 6,766 + 5,372$

2. $97,611 + 46,844 + 83,758 \bigcirc 95,897 + 47,991 + 83,742$

3. $145,877 + 292,145 + 349,522 \bigcirc 144,361 + 293,420 + 349,763$

4. $1,543,987 + 2,676,234 + 3,433,000 \bigcirc 1,542,855 + 2,700,377 + 3,336,791$

**C** For some of the problems below, the sums are not correct. Use estimation to check the reasonableness of the sums. Then correct the addition and write the correct answer below the incorrect ones.

| 1. | | 2. | | 3. | | 4. | |
|---|---|---|---|---|---|---|---|
| | 1,876,754 | | 98,675 | | 567,983 | | 767,900,000 |
| | 2,634,771 | | 86,564 | | 432,654 | | 533,800,000 |
| | + 4,382,573 | | 45,761 | | + 299,435 | | + 485,600,000 |
| | 7,792,998 | | + 38,758 | | 1,300,072 | | 1,675,300,000 |
| | | | 249,648 | | | | |

_____    _____    _____    _____

# ADDING INTEGERS

**Integers are the set of whole numbers and their opposites.**

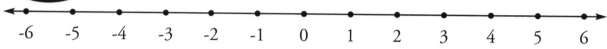

-6  -5  -4  -3  -2  -1  0  1  2  3  4  5  6

To add integers with the same sign, remember:

• the sum of two positive integers is positive    $4 + 3 = 7$

• the sum of two negative integers is negative.    $-6 + -3 = -9$

The absolute value of an integer is its distance from zero on a number line. | -14 | means "the absolute value of minus 14."

$| -14 | = 14$       $| 7 | = 7$

To add integers with different signs, remember:

• subtract their absolute values, and use the sign of the integer with the greater absolute value

$7 + -14 =$

$| -14 | - | 7 | = -7$

**A** Find each absolute value.

1. | 5 |  _____

2. | -16 |  _____

3. | 13 |  _____

4. | 21 |  _____

5. | - 9 |  _____

6. | - 30 |  _____

7. | 25 |  _____

8. | - 5 |  _____

9. | - 28 |  _____

10. | - 34 |  _____

**B** Add.

1. 2 + -7 =

2. -4 + 3 =

3. 5 + -6 =

4. 9 + -7 =

5. 9 + -8 =

6. -6 + 4 =

7. -3 + -18 =

8. -14 + -5 =

9. -6 + -11 =

10. -5 + -15 =

11. -13 + -8 =

12. - 9 + - 10 =

13. 15 + -72 =

14. -40 + 35 =

15. 22 + -48 =

16. 48 + -82 =

17. 15 + -78 =

18. -14 + 20 =

19. -32 + -42 =

20. - 67 + -14 =

21. 32 + -42 =

22. - 23 + -54 =

23. - 18 + -42 =

24. - 56 + -31 =

-100 -90 -80 -70 -60 -50 -40 -30 -20 -10 -0 10 20 30 40 50 60 70 80 90 100

# SOLVING WORD PROBLEMS

**Problem Solving Steps**
1. *Understand* the question.
2. Find the important *facts*.
3. *Solve* the problem.
4. *Check* your answer.

**Problem Solving Strategy**

**Look for a pattern.**

Tom walks one mile on Tuesday, three miles on Wednesday, and five miles on Thursday. If he continues at this rate, how far will he walk on Friday?

**Pattern:** Each day, Tom walks two miles more than he did the day before. Therefore, he will walk seven miles on Friday.

**Problem Solving Strategy**

**Draw a diagram or make a list.**

Jean, Barbara, Sue, and Mary are sitting in a row. Jean sits next to Barbara. Sue sits between Mary and Barbara. Mary sits at one end of the row. What order are they seated in?

**List:** Mary, Sue, Barbara, and Jean.

 **A** Answer the questions using the chart. Use the problem solving strategies.

1. Videos just went on sale at Ann's Video Heaven. Jane likes *Stuart Little* and *Titanic*. On Saturday, she went to Ann's to buy the videos. If she buys both of them, how much will she spend?

| Videos for Sale | |
|---|---|
| **Title** | Price |
| **The World is Not Enough** | $19.98 |
| **Stuart Little** | $24.96 |
| **Star Wars Phantom Menace** | $24.98 |
| **Titanic** | $14.98 |
| **Pochahontas** | $29.99 |

*Source: Videoscan*

2. Tony went to Ann's Video Heaven to buy a video. He decided to buy all five videos on the chart. How much money did Tony spend?

3. Judy discovered that she could only afford the three least expensive videos. How much money does Judy need to bring with her?

4. Tom lives four miles from Ann's Video Heaven. He starts walking at 1:00 PM and arrives at Ann's at 2:30 PM. He wants to buy three copies of *Stuart Little* for his nieces and nephews. When he counts his change, he finds that he has $5.00. How much did he spend for the three videos?

**B** Answer the questions using the chart. Use the problem solving strategies.

1. The high temperature in Seattle occurred at 5:00 PM. The temperature drops 2 degrees per hour starting at 6 PM. What was the temperature at 8 PM?

| High Temperatures for July 9 | |
|---|---|
| City | High Temperature |
| Chicago | 88°F |
| Denver | 93°F |
| Los Angeles | 79°F |
| Seattle | 67°F |
| New York | 78°F |

*Source: Newsday 2000 Weathernews*

2. The temperature in New York was 60°F at 7 AM. The temperature rises at a rate of 3 degrees per hour. What time would it be when the high temperature of 78°F is reached?

3. Jim wants to see the high temperatures for Phoenix, Omaha, Juneau, and Buffalo. Buffalo's high temperature is between the high temperatures of Phoenix and Omaha. Phoenix has the hottest high temperature. Juneau has the coolest high temperature. Arrange the cities in order of their temperatures from coolest to hottest.

**TEST TRACK**

**A** Respond to these items in your own words.

1. Write your own word problem using the following number sentence:
347 + 239 + 875 = 1,461

_____

_____

_____

_____

2. What property is demonstrated by 67,587 + 0 = 67,587? _____

3. Explain why the number 34,688 does not round to 34,000.

_____

_____

4. Three sisters want to buy a CD player. Sharon and Alaina each have $60.00. Their sister Katie has $75.00. How much money do they have in all? Show your work.

5. Last week, 55,822 fans paid to attend the football game. 3,197 fans had free passes to the game. How many fans attended the game? Show how you can use estimation to check the reasonableness of your answer.

6. Use the numbers 548, 879, and 331 to write a number sentence that demonstrates the commutative property.

**B** Solve. Show your work.

1. Jim bicycled 37 miles on Tuesday. He spent $16.00 on meals. On Wednesday, he rode 23 miles and spent $22.00. On Thursday, he spent $31.00 and rode 29 miles. How many miles did he travel in all?

2. Manuela's town set up a paper recycling program. In the first month, the town collected 45,672 pounds of paper. The next month, they collected 23,986 pounds of paper. In the third month, they collected 56,113 pounds of paper. The month after that, another 34,866 pounds of paper was collected for recycling. Is 160,637 a **likely** total for the amount of paper collected during the four months? Show how you used estimation to answer the question.

## ADDING FRACTIONS

When fractions have the same denominators, add their numerators to find their sum.

numerator
denominator     $\frac{1}{4} + \frac{1}{4} + \frac{2}{4} = \frac{4}{4}$

**When fractions have different denominators, find their least common denominator. Convert the fractions and add.**

$$\frac{1}{2} = \frac{3}{6}$$
$$+ \frac{1}{3} = \frac{2}{6}$$
$$\overline{\phantom{xxx} \frac{5}{6}}$$

| R E M E M B E R |
|---|
| The least common denominator of two fractions (LCD) is the least common multiple of their denominators.<br>Multiples of 2: 2, 4, **6**<br>Multiples of 3: 3, **6**<br>The LCD of $\frac{1}{2}$ and $\frac{1}{3}$ is 6. |

 **Answer the questions using the chart below.**

Melissa helps her friends build a birdhouse. She has a set of six drill bits to use with the drill. The size of each drill bit appears on the chart.

| Size (in inches) | $\frac{1}{16}$ | $\frac{1}{4}$ | $\frac{1}{8}$ | $\frac{3}{16}$ | $\frac{7}{64}$ | $\frac{3}{32}$ |
|---|---|---|---|---|---|---|
| **Drill Bit Number** | 1 | 2 | 3 | 4 | 5 | 6 |

1. Does the Number 4 drill bit make a smaller hole than the Number 2 drill bit? Show your work.

2. List the drill bits in order by size from largest to smallest.

**B** Add the fractions. Show your work.

1. $\dfrac{1}{6} + \dfrac{2}{5} =$

2. $\dfrac{1}{5} + \dfrac{2}{8} =$

3. $\dfrac{1}{4} + \dfrac{1}{3} =$

4. $\dfrac{3}{5} + \dfrac{1}{7} =$

5. $\dfrac{1}{3} + \dfrac{1}{4} + \dfrac{1}{6} =$

**C** Write a number sentence for each problem and solve.

1. Terri cuts two pieces of cake for her brother. The first piece was $\dfrac{1}{6}$ of the cake. The second piece was $\dfrac{1}{3}$ of the cake. How much of the chocolate cake did Terri give her brother?

2. Emmanuel reads $\dfrac{2}{5}$ of a book on Monday, and $\dfrac{1}{3}$ on Tuesday. How much of the book did Emmanuel read in two days?

3. Joan painted $\dfrac{1}{4}$ of her room the first day and $\dfrac{1}{3}$ the next day. Did she paint more than $\dfrac{1}{2}$ of her room? Explain your answer.

# ADDING MIXED NUMBERS

**Follow these steps to add mixed numbers.**

$$29\frac{2}{5} + 34\frac{4}{5} =$$

- add the whole numbers $\quad 29 + 34 = 63$

- add the fractions $\quad \frac{2}{5} + \frac{4}{5} = \frac{6}{5}$

- change improper fraction $\quad \frac{6}{5} = 1\frac{1}{5}$
  to a mixed number

- add $\quad 63 + 1\frac{1}{5} = 64\frac{1}{5}$

**Checking Strategy**

$$33\frac{1}{5} + 17\frac{3}{4} = 50\frac{19}{20}$$

- Use estimation to check your results.

- Round the whole numbers $\quad 30 + 20$

- Round fractions to 1 or 0 $\quad \dfrac{0 + 1}{}$

- add and compare $\quad 30 + 21 = 51 \;$ is close to $50\frac{19}{20}$

 **A** Use the chart showing stock prices to answer the questions. Show your work.

1. If the price per share of America Online stock increases by $\frac{1}{2}$, what will the new price be?

| Today's Stock Prices | |
|---|---|
| **Stock** | Price Per Share |
| **America Online** | $55\frac{3}{4}$ |
| **Yahoo!** | $117\frac{7}{16}$ |
| **Microsoft** | $79\frac{7}{8}$ |
| **McDonalds** | $32\frac{13}{16}$ |
| **Sony** | $104\frac{3}{8}$ |
| **Disney** | $38\frac{1}{4}$ |

2. Silvia expects the price per share of McDonalds stock to rise by $1\frac{5}{8}$. What will the new price per share be for McDonalds?

3. If the price per share of Microsoft increases $\frac{3}{16}$, what will the new price be?

4. A new company has a price per share that is $2\frac{7}{8}$ higher than the price per share of Disney. What is the price per share of the new company?

**B**  Solve. Show your work.

1. On her diet, Margaret lost $1\frac{1}{4}$ pounds the first week, $2\frac{1}{2}$ pounds the second week, and $1\frac{1}{2}$ pounds the third week. How much weight did she lose?

2. In December, the snowfall was $3\frac{3}{4}$ feet. In January, the snowfall was $4\frac{1}{2}$ feet. In February, the snowfall was 5 feet. What is the total snowfall during these three months?

3. Paul was training for the big race. The first day he ran $2\frac{2}{5}$ miles. On the second day, he covered $3\frac{1}{2}$ miles. On the third day, he completed $4\frac{1}{4}$ miles. On the fourth day, he ran $3\frac{1}{4}$ miles. How far did Paul run?

**C**  Add. Use estimation to check your answers.

1.  $17\frac{1}{2}$
    $+\ 34\frac{3}{4}$
    _____

2.  $2\frac{1}{5}$
    $+\ 9\frac{5}{7}$
    _____

# ADDING DECIMALS AND PERCENTS

**Remember**
- **A number written as a percent means *parts per hundred*.**
35% = 35 parts out of 100.

- **For every percent, there is an equivalent decimal.**
35% = 0.35    3% = 0.03

- **Be sure to line up the decimal points when adding:**

$$
\begin{array}{r}
34.76 \\
31.90 \\
+\ \ 4.83 \\
\hline
71.49
\end{array}
$$

**Checking Strategy**
- **Add the numbers again in reverse order**
34.76 + 31.9 + 4.83  = 71.49
4.83  + 31.9 + 34.76 = 71.49

 Use the charts to answer the questions. Show your work.

1. Is a player who makes 0.70 of his free-throws a better free-throw shooter than Kevin Garnett?

| Career Free-Throw Percentages | |
|---|---|
| **Player** | Free-Throw Percentage |
| **Grant Hill** | 75% |
| **Detlef Schrempf** | 80% |
| **Kevin Garnett** | 74% |
| **Karl Malone** | 73% |
| **Vin Baker** | 62% |

2. If Vin Baker improves his free-throw shooting percentage by 13%, he will have the same percentage as another player. Who is that player?

*Source: http://www.cbs.sportsline.com*

3. What is Kevin Garnett's free-throw percentage expressed as a decimal?

4. Marsha made a chart comparing the heights of the trees in her yard. List the trees in order from the shortest to the tallest.

| Tree | Height in Feet |
|---|---|
| Sassafras | 58.85 |
| Red maple | 76.91 |
| Willow | 69.6 |
| Oak | 96.72 |

**B** Find the sums. Check your answers.

1. 15.08 + 23.7 =

2. 675.1 + 0.004 =

3. 89.57 + 24.03 =

4. 64.6 + 17.002 + 123.05 =

5. 398.14 + 0.08 + 0.65 =

**C** Write a sentence or two explaining your answer to each question.

1. Is 0.79% of a number equal to 79% of that same number?

_____

_____

2. If you score 87% on a 100 question math test, have you answered more than 80 questions correctly? Explain.

_____

_____

3. Is 56% of a number greater than 0.65 of that same number?

_____

_____

# MIXED PRACTICE

 **A** Write each in simplest form.

1. $\frac{50}{60} =$

3. $\frac{42}{49} =$

5. $5\frac{9}{81} =$

2. $\frac{21}{25} =$

4. $\frac{15}{20} =$

6. $3\frac{2}{6} =$

**B** Change each fraction to a mixed number. Write in simplest form.

1. $\frac{23}{13}$

3. $\frac{18}{12}$

5. $\frac{62}{27}$

2. $\frac{36}{24}$

4. $\frac{21}{14}$

6. $\frac{90}{36}$

**C** Change each mixed number to an improper fraction.

1. $1\frac{2}{3} =$

3. $6\frac{1}{7} =$

5. $1\frac{4}{5} =$

2. $7\frac{1}{4} =$

4. $9\frac{2}{9} =$

6. $6\frac{1}{8} =$

**D** Add. Write the sum in simplest form.

1. $1\frac{1}{2} + \frac{1}{2} =$

3. $3\frac{1}{3} + 1\frac{2}{3} =$

5. $2\frac{7}{8} + 1\frac{3}{8} =$

2. $4\frac{1}{2} + 3\frac{1}{2} =$

4. $3\frac{1}{4} + 2\frac{3}{4} =$

6. $1\frac{1}{6} + \frac{5}{6} =$

**E** Add. Write the sum in simplest form.

1. $\frac{1}{4} + \frac{3}{5} =$

3. $\frac{5}{8} + \frac{1}{6} =$

5. $\frac{1}{8} + \frac{1}{6} =$

2. $\frac{5}{8} + \frac{3}{7} =$

4. $\frac{1}{7} + \frac{2}{9} =$

6. $\frac{2}{7} + \frac{3}{14} =$

**F** Add.

1. $0.3 + 0.03 =$

3. $1.62 + 2.5 =$

5. $63.4 + 0.07 =$

2. $1.67 + 15.8 =$

4. $2.217 + 0.004 =$

6. $2.3 + 8.9 =$

# TEST TRACK

 **A**  Choose the best answer to each question.

1. Percent means —

   Ⓐ part per ten                 Ⓒ part per thousand

   Ⓑ part per hundred             Ⓓ part per million

2. For every percent there is an equivalent —

   Ⓐ integer                      Ⓒ improper fraction

   Ⓑ triangle                     Ⓓ decimal

3. A good way to check addition is to —

   Ⓐ subtract absolute values     Ⓒ add numbers in reverse order

   Ⓑ find an equivalent fraction  Ⓓ line up decimal points

4. Improper fractions are —

   Ⓐ in simplest form             Ⓒ greater than 1

   Ⓑ less than 1                  Ⓓ integers

5. Which is NOT equivalent to $\frac{50}{100}$ ?

   Ⓐ $\frac{1}{2}$                Ⓒ 50%

   Ⓑ 0.05                         Ⓓ 0.5

6. To add fractions with unlike denominators —

   Ⓐ estimate their product

   Ⓑ find their greatest common factor and convert

   Ⓒ change each to an improper fraction

   Ⓓ find their least common multiple and convert

**B** Add.

1.
$$\frac{1}{2}$$
$$\frac{1}{3}$$
$$+ \frac{1}{4}$$
_____

2.
$$\frac{2}{3}$$
$$\frac{1}{3}$$
$$+ \frac{5}{6}$$
_____

3.
$$\frac{3}{4}$$
$$1\frac{1}{2}$$
$$+ 2\frac{5}{8}$$
_____

4.
$$4\frac{1}{6}$$
$$3\frac{2}{3}$$
$$+ 2\frac{5}{12}$$
_____

5.
$$1\frac{2}{7}$$
$$2\frac{3}{7}$$
$$+ 1\frac{5}{21}$$
_____

6.
$$5\frac{2}{5}$$
$$6\frac{3}{4}$$
$$+ \frac{9}{10}$$
_____

7.
    4.35
    0.03
  + 4.60
_____

8.
    1.88
   21.03
  + 4.09
_____

9.
   12.56
   43.65
  + 5.07
_____

10.
   12.70
   80.40
  + 0.04
_____

11.
   23.150
    1.700
  + 0.344
_____

12.
   24.15
    0.15
  + 5.67
_____

**C** Add.

1. 0.913 + 3.4 =

2. 1.37 + 0.08 =

3. 34.71 + 23.07 =

4. 7.534 + 8.096 =

5. 0.007 + 4.87 =

6. 68.58 + 0.925 =

# UNDERSTANDING SUBTRACTION

Here are the parts of a subtraction problem.

$$657 \leftarrow \text{Minuend}$$
$$-324 \leftarrow \text{Subtrahend}$$
$$333 \leftarrow \text{Difference}$$

Addition and subtraction are inverse operations.

16 + 5 = 21     21 - 5 = 16
5 + 16 = 21     21 - 16 = 5

Checking Strategy

Because addition and subtraction are inverse operations, opposites of each other, we can check subtraction with addition.

$$657$$
$$-324$$
$$333 \leftarrow \text{Difference}$$
$$+324 \leftarrow \text{Add Subtrahend}$$
$$657 \leftarrow \text{Original Minuend}$$

 **A** Select the best answer to each question about the chart.

| Skyscrapers of the World | | | |
|---|---|---|---|
| Building | Location | Height in Feet | Year Built |
| Rialto Towers | Melbourne | 825 | 1985 |
| One World Trade Center | New York | 1,368 | 1972 |
| First Canadian Place | Toronto | 978 | 1975 |
| Texas Commerce Tower | Houston | 1,002 | 1982 |
| Overseas Union Bank | Singapore | 919 | 1986 |

*Source: Marshall Geometta's Hot 500*

1. How much taller is First Canadian Place than Rialto Towers?

A 153 feet      B 163 feet      C 142 feet      D 152 feet

2. You have found the difference between the height of One World Trade Center and the Texas Commerce Tower. Which number sentence shows the inverse operation used to check the answer?

A  1,368 + 1002 = 366

B  1,368 + 1002 = 2,370

C  1,002 + 366 = 1,368

D  366 + 1,368 = 1,734

3. Which of the following problems cannot be solved using the information in the chart?

A  How much shorter is the Empire State Building than One World Trade Center?

B  What is the combined height of the buildings?

C  How much shorter is the Rialto Towers than First Canadian Place?

D  How long ago was the Texas Commerce Tower built?

**B**  Find the differences. Check your work using the inverse operation.

1.
```
   5,478
 - 3,241
```

3.
```
   7,843
 -   621
```

5.
```
   55,877
 - 23,456
```

2.
```
   69,836
 - 27,322
```

4.
```
   27,328
 -  6,115
```

6.
```
   343,245
 -  38,953
```

**C**  Respond to the following items using your own words.

1. Write related addition and subtraction number sentences for 10, 7, and 3.

2. Create your own word problem using the numbers 7,753 and 3,452.

# SUBTRACTING WITH REGROUPING

**Remember**
Some subtraction problems require regrouping. Here's how:

$$\begin{array}{r} 3^2 \ 4^{14} \ 4, \ 9^8 \ 3^{12} \ 8^{18} \\ - \ 5 \ \ \ 4, \ 7 \ \ 5 \ \ 9 \\ \hline 2 \ \ \ 9 \ \ 0, \ 1 \ \ 7 \ \ 9 \end{array}$$

 **A** Select the correct answers.

Using the Internet, Jay researched the distances from his home in St. Louis, Missouri to cities around the world. This chart shows what he found.

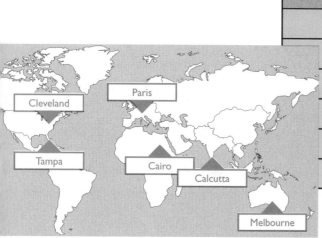

| City | Country | Distance in Miles |
| --- | --- | --- |
| Paris | France | 4,395 |
| Cleveland | U.S.A. | 494 |
| Calcutta | India | 8,222 |
| Melbourne | Australia | 9,477 |
| Tampa | U.S.A. | 861 |
| Cairo | Egypt | 6,396 |

1. How much farther is it from St. Louis to Calcutta than from St. Louis to Paris?

   A 3,827 miles      B 12,617 miles      C 4,837 miles      D 3,626 miles

2. Jay calculated the difference in mileage from St. Louis to Cairo, and then St. Louis to Tampa. What was the difference?

   A 5,535 miles      B 6,535 miles      C 7,257 miles      D 5,434 miles

3. What is the difference in the distances between the city closest to St. Louis and the city farthest away?

   A 8,983 miles      B 9,083 miles      C 9,971 miles      D 8,872 miles

**B**  Find the differences. Show your work.

1.    62,354
      -3,829

3.    892,152
      - 654,478

5.    483,198
      -75,551

2.    343,245
      -38,953

4.    54,133
      -6,876

6.    82,595
      -79,476

**C**  Answer the questions about the chart.

5 students brought school supplies.
Here's what they budgeted and spent.

| Name | Budgeted | Spent |
|---|---|---|
| Melissa | $92 | $39 |
| Bob | $92 | $96 |
| Tom | $173 | $117 |
| Chris | $87 | $38 |
| Mary | $129 | $72 |

1. Who had the most money left? _____

2. How much less did Chris have left than Mary? _____

3. How much more did Tom budget than Mary? _____

4. Mary said she had more money left than Bob. Is this true? _____

5. Who had the least money left? _____

**D**  Solve the following word problems. Show your work. Use inverse operations to check your answer.

1. The first week of the football season, 72,431 fans came to the game. The next week, 68,787 fans went to the stadium, and the third week, 71,993 fans attended. What is the difference in attendance between the first and second weeks? What is the difference in attendance between the second and third weeks?

2. Marty was reading the weather forecast for July 23rd. The forecasted high temperature was 83 degrees. There was a 30% chance of showers. Sunset would occur at 8:27 PM. At night, the humidity was expected to rise to 75% and the temperature would fall to 59 degrees. What was the difference between the forecasted high and low temperatures?

# SUBTRACTING WITH ZEROS

Every year the United States government spends money to help maintain highways. The chart below shows how many miles of highway were repaired with government funds during selected years.

How many more miles of highway received Federal help during 1994 than 1992?

$$
\begin{array}{r}
948,053 \\
- 935,101 \\
\hline
12,952
\end{array}
$$

| Federal Aid to Highways | |
|---|---|
| Year | Miles |
| 1980 | 827,956 |
| 1982 | 830,895 |
| 1985 | 843,803 |
| 1989 | 851,607 |
| 1992 | 935,101 |
| 1994 | 948,053 |

 **A**   Choose the correct answers. Use the chart above.

1. How many more miles of highway were helped by the government during 1994 than during 1980?

   A 120,097      B 120,997      C 120,197      D Answer not here.

2. How much was the increase from 1980 to 1982?

   A 2,939 miles      B Answer not here.      C 2,737 miles      D 17,939 miles

3. The difference in highway mileage between 1985 and 1989 was—

   A 8,804 miles      B 8,603 miles      C 8,604 miles      D Answer not here.

4. How many more miles of highway received government aid during 1992 than 1982?

   A 103,206 miles      B 103,296 miles      C Answer not here.      D 103,306 miles

**B** Find the differences.

1.     6,980
     -3,783

2.     305,101
     - 13,565

3.     43,000
     -28,146

4.     19,080
     -14,795

5. 830,453 - 247,867 =

6. 306,890 - 199,811 =

7. 23,087 - 1,757 =

8. 564,310 - 49,184 =

**WORK SPACE**

**C** Solve these problems and check with addition. Show your work.

1. If Carol read 100 pages Monday night and 79 pages Tuesday night, how many more pages did she read on Monday than Tuesday?

2. The population of Northville is 16,090, while Albertson has 14,226 residents. What is the difference in the populations of the two towns?

3. Brian has $190 to spend for presents. If he spends $98, will he have at least $90 left?

4. A baseball player hit 580 home runs in his career, while another player hit 393 home runs. Is the difference in their career home run totals greater than 190?

# SUBTRACTING INTEGERS

**To subtract an integer, switch its sign and add.**

$$
\begin{array}{r}
-13 \\
- \quad +2 \\
\hline
\end{array}
\qquad
\begin{array}{r}
-13 \\
+ \quad -2 \\
\hline
-15
\end{array}
\quad
\begin{array}{l}
\leftarrow \text{minuend} \\
\leftarrow \text{subtrahend} \\
\leftarrow \text{difference}
\end{array}
$$

$$
\begin{array}{r}
+12 \\
- \quad -16 \\
\hline
\end{array}
\qquad
\begin{array}{r}
+12 \\
+ \quad +16 \\
\hline
28
\end{array}
$$

$$
\begin{array}{r}
-18 \\
- \quad -4 \\
\hline
\end{array}
\qquad
\begin{array}{r}
-18 \\
+ \quad +4 \\
\hline
-14
\end{array}
$$

**A** Subtract.

1. +2 - -7 =

2. -4 - +3 =

3. +5 - -6 =

4. +9 - -6 =

5. +9 - -8 =

6. -6 - +4 =

7. -3 - -18 =

8. -12 - -5 =

9.  -6 - -11 =

10.  -5 - -16 =

11.  -19 - -8 =

12.  - 9 - -10 =

13.  +5 - -32 =

14.  -40 - +35 =

15.  +2 - -41 =

16.  +40 - -82 =

17.  +15 - -78 =

18.  -14 - +10 =

19.  -32 - -2 =

20.  - 7 - -14 =

21.  -3 - -42 =

22.  - 23 - -54 =

23.  - 8 - -42 =

24.  - 56 - -31 =

-100 -90 -80 -70 -60 -50 -40 -30 -20 -10 -0 10 20 30 40 50 60 70 80 90 100

## TEST TRACK

This table shows the number of private industry employees working from January to June in different years.

| National Employment in Private Industry | | | | |
|---|---|---|---|---|
| 1990 | 1991 | 1992 | 1993 | 1994 |
| **January** 90,908,000 | 90,441,000 | 89,561,000 | 90,775,000 | 93,328,000 |
| **February** 91,190,000 | 90,168,000 | 89,527,000 | 91,060,000 | 93,544,000 |
| **March** 91,259,000 | 89,984,000 | 89,537,000 | 90,994,000 | 93,961,000 |
| **April** 91,209,000 | 89,779,000 | 89,702,000 | 91,284,000 | 94,292,000 |
| **May** 91,261,000 | 89,735,000 | 89,871,000 | 91,592,000 | 94,597,000 |
| **June** 91,312,000 | 89,765,000 | 89,915,000 | 91,723,000 | 94,886,000 |

*Source: Bureau of Labor Statistics*

**A** Use your calculator to find the difference in the totals between:

1. January and February, 1990 and March and April, 1992

    Ⓐ 2,859,000      Ⓑ 182,098,000      Ⓒ 179,239,000      Ⓓ 5,359,000

2. May and June, 1991 and April and May, 1994

    Ⓐ 9,389,000      Ⓑ 179,500,000      Ⓒ 188,889,000      Ⓓ 8,868,900

3. February and April, 1993 and April and June, 1990

    Ⓐ 177,000      Ⓑ 2,344,000      Ⓒ 2,521,000      Ⓓ 263,000

4. March and June, 1994 and January and May, 1992

    Ⓐ 9,415,000      Ⓑ 188,847,000      Ⓒ 179,432,000      Ⓓ 8,532,000

**B** Answer the following questions or respond using your own words.

1. How are the numbers 2, 7, and 5 related?

2. In the subtraction problem below, which number is the minuend?

   7,345
   -2,658

3. Could 675,341 - 78,923 be checked by adding the difference to 675,341? Solve the subtraction problem and show your checking.

4. Does 50,302 - 16,456 = 33,846? Show your work.

5. Is 68,455 + (645 - 9,700) > 68,877 - 9,312? Show your work

6. Last year, 1,530 students attended Northside Junior High. This year, 1,779 students were in school. What is the difference in attendance between the two years?

7. How the numbers 25, 10, and 20 related?

8. Sean found that 47,345 more than 61,978 is 14,633. What error did he make?

9. In a subtraction problem, is the subtrahend usually smaller than the minuend? Give an example.

10. How are the numbers 75, 26, and 49 related?

11. Solve and check: 60,810 - 53,714

12. Estimate the difference between 433,101 and 24,657. Solve using the estimate, as well as the actual numbers.

## SUBTRACTING FRACTIONS

**When fractions have the same denominator, *subtract* the numerators and leave the denominators the same.**

$\frac{4}{5} - \frac{3}{5} = \frac{1}{5}$    **subtract** the numerators
denominators remain the same

**When fractions have different denominators, find the *least common denominator*.**

multiples of 4 are 4, 8, 12, 16, **20**
multiples of 5 are 5, 10, 15, **20**
**least common denominator** is 20

| | Convert: | Convert: | Subtract: | Answer: |
|---|---|---|---|---|
| $\frac{4}{5} - \frac{1}{4} = ?$ | $\frac{4}{5} \times \frac{4}{4} = \frac{16}{20}$ | $\frac{1}{4} \times \frac{5}{5} = \frac{5}{20}$ | $\frac{16}{20} - \frac{5}{20} = \frac{11}{20}$ | $\frac{4}{5} - \frac{1}{4} = \frac{11}{20}$ |

 **A**   Read the recipe below. Answer the questions. Show your work.

| Excellent Oatmeal Raisin Muffins | | | |
|---|---|---|---|
| Ingredient | Quantity | Ingredient | Quantity |
| Flour | $\frac{1}{2}$ Cup | Sugar | $\frac{1}{4}$ Cup |
| Baking Powder | 1 Teaspoon | Egg White | $\frac{1}{4}$ Cup |
| Cinnamon | $\frac{1}{2}$ Teaspoon | Milk | $\frac{1}{2}$ Cup |
| Nutmeg | $\frac{1}{4}$ Teaspoon | Raisins | $\frac{1}{2}$ Cup |
| Salt | $\frac{1}{8}$ Teaspoon | Oatmeal | $2\frac{1}{4}$ Ounces |
| Margarine | $\frac{3}{8}$ Cup | | |

1. Increase the amount of flour in the recipe to $\frac{7}{8}$ of a cup. How much **more** flour is needed?

2. Mark uses $\frac{1}{8}$ cup **less** of margarine in his recipe. How much margarine does he use?

3. What is the least common denominator used to find the difference between the amount of margarine and raisins?

4. How much less salt than nutmeg is needed?

**B** Solve the problems.

1. $\dfrac{7}{8} - \dfrac{3}{8} =$

3. $\dfrac{2}{3} - \dfrac{1}{5} =$

2. $\dfrac{8}{9} - \dfrac{4}{9} =$

4. $\dfrac{6}{7} - \dfrac{1}{2} =$

**C** Solve. Show your work.

1. If it rained $\dfrac{1}{4}$ of an inch yesterday and $\dfrac{3}{4}$ of an inch today, how much more rain fell today?

2. Mike drew two lines $\dfrac{3}{4}$ of an inch apart. If he draws them $\dfrac{1}{2}$ an inch closer, how far apart will the lines be?

3. Sam ate $\dfrac{7}{8}$ of his medium cheese pizza. Joe ate $\dfrac{5}{8}$ of his medium pepperoni pizza. How much more pizza did Sam eat than Joe?

4. Linda has two piles of papers on her desk. One pile is $\dfrac{13}{16}$ inches high. The other pile is $\dfrac{7}{16}$ inches high. How much taller is one pile than the other pile?

5. It is $\dfrac{7}{10}$ of a mile from Bill's house to Fran's house. Juan's house is $\dfrac{1}{5}$ of a mile closer to Bill's house than it is to Fran's house. How far is it from Juan's house to Bill's house?

# SUBTRACTING MIXED NUMBERS

Sometimes mixed numbers need to be regrouped before subtracting.

**with common denominators:**

$6\frac{1}{3} - 4\frac{2}{3} =$

$6\frac{1}{3} = 5\frac{1}{3} + \frac{3}{3} = 5\frac{4}{3}$

$5\frac{4}{3} - 4\frac{2}{3} = 1\frac{2}{3}$

**with different denominators:**

$7\frac{3}{8} - 3\frac{1}{2} =$

$7\frac{3}{8}$    multiples of 8: **8**, 16, 24

$-3\frac{1}{2}$    multiples of 2: 2, 4, 6, **8**
The LCD is 8

$7\frac{3}{8} = 6\frac{3}{8} + \frac{8}{8} = 6\frac{11}{8}$

$6\frac{11}{8} - 3\frac{4}{8} = 3\frac{7}{8}$

$7\frac{3}{8} - 3\frac{1}{2} = 3\frac{7}{8}$

 **A**    Look at the chart below to solve each word problem. Show your work.

Cheryl is interested in dinosaurs. She visits the Natural History Museum and sees models of dinosaurs.

| Dinosaur | Approximate Length |
|---|---|
| **Albertosaurus** | $28\frac{3}{8}$ feet |
| **Giganotosaurus** | $46\frac{3}{8}$ feet |
| **Spinosaurus** | $5\frac{1}{4}$ feet |
| **Apatosaurus** | $73\frac{2}{3}$ feet |
| **Lesothosaurus** | $3\frac{1}{2}$ feet |
| **Plateosaurus** | $26\frac{1}{3}$ feet |
| **Hypacrosaurus** | $31\frac{3}{4}$ feet |

1. How much shorter is the Spinosaurus than the Hypacrosaurus?

2. How much longer is the Apatosaurus than the Plateosaurus?

3. What is the difference in length between the Hypacrosaurus and the Lesothosaurus?

4. How much shorter is the Lesothosaurus than the Plateosaurus?

 **B** Solve.

1. $34\frac{1}{8} - 15\frac{3}{8} =$

3. $12\frac{3}{4} - 2\frac{7}{8} =$

2. $7\frac{4}{9} - 3\frac{1}{2} =$

4. $13\frac{3}{7} - 4\frac{1}{2} =$

**C** Use the table below to answer the questions. Show your work.

| Frank's Mileage Log | |
|---|---|
| Day | Miles Ridden |
| Sunday | $15\frac{7}{10}$ |
| Monday | $6\frac{1}{2}$ |
| Tuesday | $10\frac{9}{10}$ |
| Wednesday | $11\frac{3}{4}$ |
| Thursday | $8\frac{3}{5}$ |
| Friday | $13\frac{1}{10}$ |
| Saturday | $12\frac{1}{4}$ |

1. How much farther did Frank ride on Sunday than Saturday?

2. What is the difference in mileage between the day with the most mileage, and the day with the least?

3. On Friday, Bob rode $2\frac{3}{4}$ fewer miles than Frank. How far did Bob ride on Friday?

4. How much shorter was Thursday's ride than Wednesday's ride?

# SUBTRACTING DECIMALS AND PERCENTS

**When you subtract decimals, line up the decimal points just as you do when you add them.**

Jean has $60.00 to buy a pair of sneakers that cost $58.25. How much change will she receive?

$60.00
$58.25
$ 1.75  change

Shoes at the Flying Foot are 25% off on Mondays. If you use the new Flying Foot Charge Card, you can get an additional 5% discount. What percent of the original price will you pay if you use your Flying Foot Charge Card on Monday?

| | | |
|---|---|---|
| 100% | | 1.00 |
| - 25% | | - .25 |
| 75% | or | .75 |
| - 5% | | - .05 |
| 70% | | .70 |

Answer: 70%

 **A**  Use the chart below to answer the questions. Show your work.

The *Just in Style* shoe store is having a big sale.  Here are the prices.

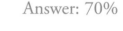

| Item | Price |
|---|---|
| **Boots** | $69.99 |
| **Boat shoes** | $59.50 |
| **Sandals** | $34.75 |
| **Loafers** | $73.99 |
| **Sneakers** | $58.25 |
| **Dress Shoes** | $119.75 |
| **Classics** | $85.25 |

1. How much change will Ken receive if he buys a pair of Classics with a $100.00 bill?

2. Melanie has $40.00 to buy a pair of sandals.  How much money will she have left over?

3. How much more money does a pair of loafers cost than boat shoes?

4. What is the difference in price between boots and dress shoes?

**B** Find the difference.

1. 728.67 - 43.99 =

2. 41.09 - 3.3 =

3. 14.23 - 0.06 =

4. 5.11 less than 63.2 =

5. 1.22 less than 7 =

6. 589.36 less than 1,098.8 =

**C** Solve. Show your work.

1. Mary Ellen saved $100.00 from her summer job. She spent 63% of her money. How much did she have left?

2. Beth had $10.00 in her pocket. She paid $6.75 to go to the movies. How much did she have left for refreshments?

3. There are 100 students in the 8th Grade. 45% of them will not go on the class trip. How many students will go on the class trip?

4. Pete wants to buy a CD that costs $16.59. He has $13.75. How much more money does he need?

5. 28% of 100 adults favor a longer school year. How many adults do not favor a longer school year?

# MIXED PRACTICE

 **Subtract. Simplify if necessary.**

1. $\dfrac{8}{9} - \dfrac{4}{9} =$

3. $\dfrac{7}{9} - \dfrac{25}{45} =$

5. $\dfrac{6}{7} - \dfrac{19}{28} =$

2. $\dfrac{9}{10} - \dfrac{3}{5} =$

4. $\dfrac{6}{11} - \dfrac{4}{33} =$

6. $\dfrac{1}{2} - \dfrac{5}{12} =$

**B** **Subtract. Simplify if necessary.**

1. $\dfrac{3}{4} - \dfrac{2}{3} =$

3. $\dfrac{2}{3} - \dfrac{1}{7} =$

5. $\dfrac{5}{6} - \dfrac{4}{9} =$

2. $\dfrac{4}{5} - \dfrac{1}{4} =$

4. $\dfrac{3}{5} - \dfrac{1}{8} =$

6. $\dfrac{6}{7} - \dfrac{4}{9} =$

**C** **Subtract. Simplify if necessary.**

1. $3\dfrac{7}{9} - 1\dfrac{4}{9} =$

3. $6\dfrac{7}{9} - 2\dfrac{7}{21} =$

5. $9\dfrac{6}{7} - 2\dfrac{19}{21} =$

2. $5\dfrac{9}{10} - 4\dfrac{7}{10} =$

4. $7\dfrac{7}{11} - 4\dfrac{13}{33} =$

6. $34\dfrac{1}{2} - 24\dfrac{5}{12} =$

**D** Subtract. Simplify if necessary.

1. $2\frac{3}{4} - \frac{1}{3} =$

3. $4\frac{2}{3} - 2\frac{6}{7} =$

5. $2\frac{5}{6} - \frac{5}{9} =$

2. $3\frac{3}{5} - 1\frac{1}{4} =$

4. $3\frac{1}{5} - 3\frac{1}{8} =$

6. $2\frac{3}{7} - 1\frac{4}{9} =$

**E** Subtract.

1. $168.3 - 3.6 =$

3. $127.4 - 4.9 =$

5. $4202.5 - 326.81 =$

2. $56.25 - 7.5 =$

4. $556.8 - 8.76 =$

6. $6678.9 - 593.23 =$

**F** Subtract.

1. $414.75 - 5.05 =$

3. $1615.03 - 615.36 =$

5. $980.64 - 9.08 =$

2. $20.15 - 12.06 =$

4. $171.008 - 14.806 =$

6. $662.04 - 584.003$

# SOLVING WORD PROBLEMS

**Problem Solving Strategy**
**Guess and check**

Sarah saves $3.00 more than Trevor. Don saves $10.00 more than Sarah, and Neal saves $5.00 less than Don. If they save $104.00 together, how much does each one save?

1. Find the known information.
   • total savings are $104.00

2. List each piece of information.
   • Sarah saved $3.00 more than Trevor
   • Don saved $10.00 more than Sarah
   • Neal saved $5.00 less than Don.
   • Trevor – no information

3. Make a guess: Trevor saves $15.00.

4. Check your guess.

   | | |
   |---|---|
   | Trevor | $15.00 |
   | Sarah | $18.00 |
   | Don | $28.00 |
   | Neal | $23.00 |
   | Total: | $84.00 |

   **Answer is too low.**

5. Guess again: Trevor saves $20.00.

   | | |
   |---|---|
   | Trevor | $20.00 |
   | Sarah | $23.00 |
   | Don | $33.00 |
   | Neal | $28.00 |
   | Total: | $104.00 |

   **Correct answer.**

---

**A** Use the chart below to answer the questions. Show your work.

There are many lighthouses on Long Island, near New York City. Here is a chart showing some of them.

| Lighthouse | Approximate Height |
|---|---|
| **Old Field Point** | $34 \frac{3}{4}$ feet |
| **Stepping Stones** | $46 \frac{1}{2}$ feet |
| **Montauk Point** | $110 \frac{7}{8}$ feet |
| **Fire Island** | $168 \frac{5}{8}$ feet |
| **Horton Point** | $58 \frac{1}{3}$ feet |

*Source: Http://www.longislandlighthouses.com*

1. What is the difference in height between the tallest lighthouse and the shortest lighthouse?

2. If the Stepping Stones lighthouse was $12\frac{2}{3}$ feet taller, how tall would it be?

3. If all of the lighthouses on the chart were destroyed by storms, what is the **combined** height of all the lighthouses that would need to be replaced?

**B** Solve. Show your work.

1. Dawn had $17.30 after she returned from shopping. She bought a book for $12.95, greeting cards for $6.25, a drink for $2.50, and a belt for $21.00. How much money did Dawn have before she went shopping?

2. There are 6 more students in Maria's class than in Ted's class. Tony's class has 5 fewer students than Maria's class. Tamara's class has 3 more children than Tony's class. There are 107 students in the four classes. How many students are in each class. Show your work using **guess and check**.

3. At the supermarket, Fred buys English Muffins for $1.39, chicken for $3.99, ice cream for $4.29, and cheese for $2.09. How much change will Fred get back if he pays with a $20 bill?

4. In the school bowling league, Cindy's score is 18 points more than Amy's score. If the two girl's score a total of 288 points, what was each girl's score?

# TEST TRACK

 **A** Choose the answer.

1. $\frac{7}{8} - \frac{2}{3} =$  
   Ⓐ $\frac{1}{6}$  Ⓑ $\frac{1}{8}$  Ⓒ $\frac{5}{12}$  Ⓓ $\frac{5}{24}$

2. $3\frac{1}{3} - 2\frac{1}{2} =$  
   Ⓐ $\frac{5}{6}$  Ⓑ $\frac{1}{2}$  Ⓒ $1\frac{5}{6}$  Ⓓ $1\frac{1}{3}$

3. $\frac{5}{6} - \frac{1}{4} =$  
   Ⓐ $\frac{1}{3}$  Ⓑ $\frac{1}{4}$  Ⓒ $\frac{5}{12}$  Ⓓ $\frac{7}{12}$

4. $6\frac{4}{5} - 2\frac{1}{3} =$  
   Ⓐ $4\frac{7}{15}$  Ⓑ $4\frac{5}{15}$  Ⓒ $4\frac{1}{5}$  Ⓓ $4\frac{1}{3}$

5. $2\frac{1}{3} - 1\frac{5}{6} =$  
   Ⓐ $\frac{1}{3}$  Ⓑ $\frac{2}{3}$  Ⓒ $\frac{5}{6}$  Ⓓ $\frac{1}{2}$

6. $10 - 6\frac{1}{3} =$  
   Ⓐ $4\frac{1}{3}$  Ⓑ $3\frac{2}{3}$  Ⓒ $4\frac{2}{3}$  Ⓓ $4\frac{1}{2}$

7. $8\frac{1}{4} - 3\frac{5}{6} =$  
   Ⓐ $5\frac{5}{12}$  Ⓑ $5\frac{7}{12}$  Ⓒ $4\frac{5}{12}$  Ⓓ $4\frac{7}{12}$

8. $4\frac{2}{5} - 3\frac{1}{2} =$  
   Ⓐ $4/5$  Ⓑ $\frac{9}{10}$  Ⓒ $1\frac{3}{5}$  Ⓓ $1\frac{2}{3}$

9. $7 - 3\frac{2}{3} =$  
   Ⓐ $3\ 2/3$  Ⓑ $3\frac{1}{2}$  Ⓒ $4\frac{1}{3}$  Ⓓ $3\frac{1}{3}$

10. $4\frac{7}{8} - 2\frac{5}{6} =$  
    Ⓐ $2\frac{1}{24}$  Ⓑ $1\frac{1}{3}$  Ⓒ $1\frac{1}{2}$  Ⓓ $1\frac{1}{6}$

**B** Find the differences.

1. 21.432 - 1.2 =

2. 4.32 - 0.04 =

3. 81.4 - 4.404 =

4. 54.75 - 6.214 =

5. 0.45 - 0.224 =

6. 30.9 - 5.21 =

**C** Solve. Show your work.

1. Of the 70 members of the club, only 35% attended the meeting. What percent of the membership did not attend?

2. Fran was $4\frac{1}{2}$ feet tall when she was 11 years old. By the time she was 14, she was $5\frac{3}{4}$ feet tall. How much did she grow between the ages of 11 and 14?

3. Jim paid $13.42 for a book. He paid with a $20 bill. How much change did he receive?

4. Gwen saved a $4\frac{7}{8}$ inch board from a piece of lumber that was 8 feet long. How long was the other piece?

5. Joanne and three friends sold tickets to the big game. Jim sold $\frac{1}{4}$ of them, Judy sold $\frac{1}{3}$ of them, and Jane sold $\frac{1}{6}$ of them. What fraction of the total number of tickets did Joanne sell?

# BASIC CONCEPTS OF MULTIPLICATION

**Numbers can have several factors. For example:**
1, 2, 3, 4, 6, 8, 12, and 24 are all factors of 24.

| | |
|---|---|
| 6 x 4 = 24 | 8 x 3 = 24 |
| 24 x 1 = 24 | 12 x 2 = 24 |

**We can use the distributive property to make multiplying easier.**

37 x 6 =

(30 x 6) + (7 x 6) =

180 + 42 = 222

**Understanding Multiples of 10 can help you multiply mentally.**

| | |
|---|---|
| 6 x 20 = | 120 |
| 6 x 200 = | 1,200 |
| 6 x 2000 = | 12,000 |

| **Problem Solving Strategy** |
|---|
| Use the distributive property and your knowledge of multiples of 10 to solve problems mentally. |

**A** Look at the chart below to answer the questions. Show your work.

This chart shows the average number of students in the classrooms of seven U.S. states.

1. There are 9 classrooms at McKinley School in Erno, California. About how many students attend McKinley School?

| State | Average Number of Students Per Classroom |
|---|---|
| Alaska | 22 |
| California | 28 |
| Kansas | 21 |
| Minnesota | 25 |
| South Dakota | 20 |
| Vermont | 19 |
| Wisconsin | 23 |

*Source: U.S Department of Education*

2. About how many students attend a Wisconsin school with 10 classrooms?

3. About how many students attend a Minnesota school with 12 classrooms?

**B** Solve using the distributive property.

1. 8 x 27 =

2. 38 x 5 =

3. 19 x 6 =

4. 4 x 44 =

5. 7 x 54 =

6. 2 x 17 =

7. 8 x 43 =

8. 21 x 5 =

**C** Follow the instructions.  Show your work.

1. Write the factors of 15. _____

2. A way to write 3 x 25 = using the distributive property is _____

3. Write the number sentence below another way.

(30 x 2) + (7 x 2) =

4. The product of 7 and 40 is _____

5. Show how to use the distributive property to solve  23 x 9 =.

# MULTIPLYING BY A 1- OR 2-DIGIT NUMBER

**Multiplication Strategy**
**Use an alternative algorithm to find the answer.**

345 x 5 =

5 x 5   =   25  ← multiply the place value of each digit in the larger number by the smaller number

5 x 40  =  200

5 x 300 = 1500  ← add them together

         1725

 Use the chart to answer the questions or respond to the items that follow.

Renting a car costs more in California than in most other states because of pollution control requirements. This chart shows the average rate per day to rent various size cars in most states and in California.

| Size of Car | Daily Rate |
|---|---|
| Compact - most states | $28.00 |
| Compact - California | $31.00 |
| Mid Size - most states | $36.00 |
| Mid Size - California | $39.00 |
| Luxury - most states | $45.00 |
| Luxury - California | $47.00 |

1. Joan is traveling to Florida. She wants to rent a luxury car for a week. How much will she spend?

2. Manuel is going on vacation to California. How much will it cost him to rent a car for two weeks?

3. Tom has budgeted $500 to rent a car for two weeks in Ohio. What type of car can he rent?

4. What is the cost to rent a mid-size car in Illinois for one week?

Find the answers. Show your work.

1. 496 × 2 =

2. 4 × 812 =

3. 5 × 774 =

4. 7 × 156 =

5. 32 × 87 =

6. 15 × 688 =

7. 12 × 243 =

8. 41 × 42 =

   Use the table below to answer the questions. Show your work.

2000 students were asked to keep track of how much television they watched on a weekend day in February. Here are the results.

1. What is the total number of viewing hours for students who watch six hours of television a day?

| Number of Hours | Number of Students |
|---|---|
| Less than one hour | 116 |
| 1 | 92 |
| 2 | 347 |
| 3 | 291 |
| 4 | 522 |
| 5 | 129 |
| 6 | 143 |
| 7 | 135 |

2. Which is greater, the total viewing hours for those who watched 6 hours of television, or the total viewing hours for those who watched 3?

# BASIC CONCEPTS OF DIVISION

$$\underset{\text{dividend}}{45} \div \underset{\text{divisor}}{9} = \underset{\text{quotient}}{5}$$

**Division is the inverse operation of multiplication. Each operation can be used to check the other.**

| | | |
|---|---|---|
| 8 x 4 = 32 | 6 x 7 = 42 | 15 ÷ 5 = 3 |
| 4 x 8 = 32 | 7 x 6 = 42 | 15 ÷ 3 = 5 |
| 32 ÷ 8 = 4 | 42 ÷ 6 = 7 | 5 x 3 = 15 |
| 32 ÷ 4 = 8 | 42 ÷ 7 = 6 | 3 x 5 = 15 |

| **Be Careful** |
|---|
| 15 ÷ 3 = 5 is **not** the inverse of 15 x 3 = 45 |

**Know how to work with multiples of 10.**

| dividend | | divisor | | quotient |
|---|---|---|---|---|
| 20 | ÷ | 4 | = | 5 |
| 200 | ÷ | 40 | = | 5 |
| 2000 | ÷ | 400 | = | 5 |
| 20000 | ÷ | 4000 | = | 5 |

**A** Use the chart below to answer the questions. Show your work.

1. How many months does it take to pay for a Gas Range? Check your answer using the inverse operation.

| Appliance City | | |
|---|---|---|
| **Item** | Price | Payments Each Month |
| **Gas Range** | $792.00 | $22 |
| **Washer** | $504.00 | $21 |
| **Dryer** | $288.00 | $12 |
| **Refrigerator** | $548.00 | $18 |

2. Louise wants to buy a dryer. She can pay $24.00 a month. How many months will it take for Louise to pay for the dryer? Check your answer using the inverse operation.

3. Ralph needs a new refrigerator for his family. How many months will it take to pay for the refrigerator? Check your answer using the inverse operation.

**B**  Write three related number sentences for each item.

1. $24 \div 6 = 4$    2. $5 \times 8 = 40$    3. $6 \times 7 = 42$    4. $18 \div 3 = 6$

**C**  Respond to each item using your own words.

1. Explain why $40 \div 8$ and $400 \div 80$ are equal.

2. Why are inverse operations important?

3. Show how to find the answer to the problem $30,000 \div 5000$.

4. Show the inverse operation of $7 \times 8 = 56$.

# DIVIDING BY A 1- OR 2-DIGIT DIVISOR

**Problem Solving Strategies**
**Check for missing information.**
**Watch out for unnecessary information.**

Mrs. Lane sells used books to raise money for a class trip. She wants to earn $400.00 from the book sale. The students have 630 books. They plan to display them on five tables. They plan to sell mystery books for 50 cents and science books for 75 cents. Did the students reach their goal?

The number of tables they use to display the books is not needed to answer the question.

The number of mystery and science books is needed to answer the question.

 **Look at the chart below to answer the questions. Show your work.**

Chuck plans to travel from Toronto to several cities across Canada.

| City | Distance from Toronto in Kilometers |
|---|---|
| Calgary | 3,422 |
| Quebec | 800 |
| Saskatoon | 2,920 |
| Regina | 2,622 |
| Prince Rupert | 4,907 |

1. Chuck drove to Regina in 5 days. About how far did he go each day?

2. Chuck plans to drive 95 miles a day when he goes to Quebec. About how long will it take him to get there?

3. On his last trip, it took Chuck 11 days to reach Regina. About how far did he travel each day?

**B** Find the quotient. Show your work.

1. $803 \div 6 =$

2. $790 \div 2 =$

3. $1,040 \div 4 =$

4. $5,500 \div 20 =$

5. $600 \div 9 =$

6. $7,805 \div 13 =$

7. $6,922 \div 27 =$

8. $9,574 \div 66 =$

**C** Solve. Show your work.

1. The combined attendance for the last 11 baseball games is 429,605 fans. What is the average attendance for each game?

2. There are 1,149 students attending the Parkway Middle School. Each student travels an average of 4.2 miles to school. The principal decides that class size for next year will be 24 students per class. How many classrooms are needed for the students?

3. Michelle is taking six subjects with a different notebook for each subject. Each weeknight she averages 3.3 hours of homework and reads 75 pages in her text books. How many hours does Michelle spend on her homework each week?

# DIVIDING BY A 2- OR 3-DIGIT DIVISOR

Division and multiplication are *inverse operations.*
They can be used to check each other.

$$\underset{10,998}{\overset{\text{dividend}}{\downarrow}} \div \underset{47}{\overset{\text{divisor}}{\downarrow}} = \underset{234}{\overset{\text{quotient}}{\downarrow}}$$

Example 1: $10,998 \div 47 = 234$
Check: $234 \times 47 = 10,998$ ← Multiply the quotient by the divisor. The result should equal the dividend.

Example 2: $467 \div 21 = 22 \text{ r } 5$
Check: $22 \times 21 = 462.$
$462 + 5 = 467$ ← Add the remainder 5.

Example 3: $540 \div 90 = 6$
Check: $54 \div 9 = 6$ ← Eliminate an equal number of zeroes from the divisor and dividend.

 Use the table below to answer the questions. Show your work.

The U.S. Census Bureau predicts population growth. Here are some of its projections, in thousands for different states.

1. If Wyoming's population increases by a total of 37,240 in the next five years, how much will the population increase each year?

| Projected Population (in thousands) | | |
|---|---|---|
| State | 2005 | 2015 |
| Delaware | 793 | 851 |
| Wyoming | 559 | 636 |
| Alaska | 659 | 728 |
| North Dakota | 677 | 727 |
| Rhode Island | 986 | 989 |
| Vermont | 623 | 646 |

*Source: http://www.census.gov*

2. If we divide Alaska's population increase of 69,000 over 10 years from 2005 to 2015, the average increase is 6,900 per year. Show how to check this answer.

3. If the population in North Dakota increases by a total of 50,000 in 10 years, how much will it increase in five years if it increases at the same rate?

**B** Solve and check. Show your work.

1. 790 ÷ 25 =

2. 1,406 ÷ 36 =

3. 17,060 ÷ 651 =

4. 305,014 ÷ 232 =

**C** Follow the instructions.

1. Show whether or not (56,000 ÷ 100 = 560) and (560 ÷ 10 = 56) are equal.

2. Write a word problem that is missing some information you need to solve the problem.

3. How many zeroes could you eliminate when dividing 450,000 by 9,000? Show and check your work.

# WORKING WITH LARGE NUMBERS

You can work with large numbers by ignoring zeroes during computation and replacing them after you find the answer.

**Round to the nearest hundred million:**

$1,547,~~000,000~~

$1,547 ⟵——————— rounds to $1,500

$1,500,**000,000** ⟵—— add zeroes back in

**Use the chart below to answer the questions. Show your work.**

| Acme Sports Store Annual Sales | |
|---|---|
| Year | Sales |
| 2000 | $1,547,000,000 |
| 2001 | $1,953,600,000 |
| 2002 | $2,415,600,000 |
| 2003 | $3,021,900,000 |
| 2004 | $3,843,400,000 |

Anna wants to buy shares of Acme Sports Store stock. She checks the company's annual sales projections.

1. Each year is divided into four quarters. If sales are about equal in each quarter, estimate the sales for the first quarter of 2001 to the nearest hundred million. Show your work.

2. What is a good estimate for sales in 2002 to the nearest hundred million? Show your work.

3. Estimate sales in 2005 to the nearest hundred million if sales increase by $500 million over the previous year.

**B** Complete the number sentences and problems. Show your work.

1. 1,356,015 ÷ 17 =

2. 397 x 286 =

3. 4,780 x 376 =

4. 6,763 ÷ 21 =

**C** Follow the instructions.

1. Solve this number sentence:   29 x _____ = 45,588

2. In one school, 18 classes collected a total of $23,097 in a fund raising drive. Show how to find the approximate amount of money collected by each class.

3. There are 34 towns in the county. Each town has an average of 16,678 families. Each family has an average of two children. Write a number sentence to show how many children live in the county. Solve the problem.

4. Does 48,000,000 ÷ 6,000 have the same quotient as 480,000 ÷ 60? State your answer and prove it.

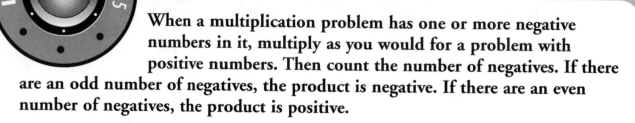

# MULTIPLYING INTEGERS

When a multiplication problem has one or more negative numbers in it, multiply as you would for a problem with positive numbers. Then count the number of negatives. If there are an odd number of negatives, the product is negative. If there are an even number of negatives, the product is positive.

+5 x -6 = -30    Here, there is only one negative (-6). Therefore, the product is negative.

-3 x -4 = +12    Here, there are two negatives (-3 and -4). Therefore, the product is positive.

| **R E M E M B E R** |
| --- |
| Positive times positive is positive. |
| Negative times positive is negative. |
| Negative times negative is positive. |

 Solve.

1. -4 x -7 =

2. +5 x - 2 =

3. -6 x -8 =

4. +9 x - 4 =

5. -7 x -7 =

6. -9 x +8 =

7. +6 x - 7 =

8. -3 x -12 =

9. +12 x -8 =

10. +31 x -8 =

13. +75 x -36 =

16. -18 x -7 =

11. -37 x -12 =

14. -65 x -21 =

17. +63 x +58 =

12. -25 x +7 =

15. -45 x +17 =

18. -74 x -17 =

**B** Solve.

1. -575 x +219 =

4. +819 x -105 =

7. +8502 x -523 =

2. +78 x -136 =

5. -720 x -286 =

8. +6349 x -358 =

3. +309 x +402 =

6. -924 x +388 =

9. -8543 x -204 =

## TEST TRACK

**A** Use the chart below to answer the questions. Show your work.

James loves learning about different animals. He learns about some North American animals on the Internet.

| Animal | Estimated Population | # of Young Born each Year |
|--------|---------------------|---------------------------|
| **American Bison** | 16,000 | 1 |
| **Bobcat** | 1,020,000 | 1 - 6 |
| **Mexican Wolf** | 34 | 2 - 8 |
| **Bald Eagle** | 110,000 | 2 |

*Source: http://www.kidsplanet.org*

1. On a recent bird census, 7300 mating pairs of Bald Eagles were counted. How many chicks will be born to the pairs next year?

2. What quotient results when you divide the population of Bobcats by 1,000?

**B** Follow the directions.

1. List the factors of 35.

2. Show how to check the result of 433 ÷ 19 =.

3. Estimate the product: 713 x 185 =.

4. In 79,870 ÷ 329, the number 329 is called the _____.

5. Show how to check the result of 43,234 x 17 =.

6. Show how to simplify the problem 36,000 ÷ 900.

 **Follow the instructions.**

1. Solve 437 x 5 using the distributive property.

2. List the factors of 48.

3. Explain why 492 x 378 = 185,976 can be checked by solving 185,976 ÷ 492 = .

4. Complete the following number sentence. Write <, >, or = in the circle.

   38,675 ÷ 221 ◯ 175 x 221

5. Show how you would estimate the quotient 674,122 ÷ 1,788 = .

## UNIT 6

# MULTIPLYING FRACTIONS

**To multiply a fraction, find the product of the numerators and the product of the denominators.**

For example:

$$\frac{4}{5} \times \frac{2}{3} = \frac{8}{15}$$

Melanie only wanted to make enough coleslaw for one person. So she only used $\frac{1}{4}$ of each ingredient called for in the recipe. How much red cabbage did she use?

**Special Creamy Coleslaw**

$\frac{1}{2}$ cup green cabbage    $\frac{1}{5}$ cup green pepper

$\frac{1}{2}$ cup red cabbage    $\frac{1}{8}$ cup cucumber

$\frac{1}{3}$ cup carrot    $\frac{1}{4}$ cup mayonnaise

$\frac{1}{4}$ cup celery    2 teaspoons sugar

$$\frac{1}{2} \times \frac{1}{4} = \frac{1}{8} \text{ cup}$$

Melanie used $\frac{1}{8}$ of a cup of red cabbage.

**A**    Use the recipe above and choose the correct answer.

1. Dan chopped $\frac{1}{2}$ cup of celery and put $\frac{1}{3}$ of it in the coleslaw. Did he use more or less than $\frac{1}{4}$ cup?

   A less      B more      C Not enough information.      D Answer not here.

2. If we cut the recipe in half, how much mayonnaise would we include in the coleslaw?

   A $\frac{1}{2}$ cup      B $\frac{1}{4}$ cup      C $\frac{3}{8}$ cup      D $\frac{1}{8}$ cup

3. If we measure out $\frac{3}{4}$ of a cup of onion but use only $\frac{1}{2}$ that amount, how much onion will we use?

   A $\frac{1}{8}$ cup      B $\frac{2}{4}$ cup      C $\frac{3}{8}$ cup      D $\frac{5}{8}$ cup

**B** Solve. Simplify if necessary.

1. $\frac{1}{3} \times \frac{1}{2} =$

2. $\frac{3}{4} \times \frac{1}{5} =$

3. $\frac{1}{2} \times \frac{5}{7} =$

4. $\frac{1}{2} \times \frac{1}{5} =$

5. $\frac{5}{6} \times \frac{3}{4} =$

6. $\frac{2}{3} \times \frac{1}{2} =$

7. $\frac{2}{5} \times \frac{1}{6} =$

8. $\frac{1}{8} \times \frac{2}{3} =$

9. $\frac{4}{10} \times \frac{2}{5} =$

> When multiplying fractions, each of the fractions being multiplied is known as a factor. In the problem, $\frac{1}{2} \times \frac{1}{5} = \frac{1}{10}$, both $\frac{1}{2}$ and $\frac{1}{5}$ are the factors, $\frac{1}{10}$ is the product.

**C** True or false? If false, write the correct answer.

1. When we multiply $\frac{2}{9} \times \frac{3}{5}$, $\frac{2}{9}$ and $\frac{3}{5}$ are the products. _____

2. Sal had $\frac{1}{3}$ of a box of cereal. He ate $\frac{3}{4}$ of that box this morning. We can say he ate $\frac{3}{12}$ or $\frac{1}{4}$ of the box of cereal. _____

3. Cynthia took $\frac{3}{4}$ of an hour to do her homework last night. Of that time, she spent $\frac{1}{2}$ on her math. She spent $\frac{5}{6}$ of an hour working on math.

_____

4. Mary walked for $\frac{1}{2}$ an hour after school yesterday. For $\frac{1}{2}$ of that time she talked on her cell phone. She talked on her cell phone for $\frac{1}{4}$ of an hour. _____

5. If the product is $\frac{2}{9}$, the factors could be $\frac{1}{3}$ and $\frac{2}{3}$. _____

# MULTIPLYING MIXED NUMBERS

**When working with a factor that is a mixed number, follow these steps:**

**Example:** $\frac{3}{4}$ x $1\frac{1}{2}$ =

**Step 1:** Convert the mixed numbers to improper fractions. $1\frac{1}{2} = \frac{3}{2}$

**Step 2:** Multiply. $\frac{3}{4}$ x $\frac{3}{2} = \frac{9}{8}$

**Step 3:** Convert the improper fraction back to a mixed number. $\frac{9}{8} = 1\frac{1}{8}$

Fred is interested in buying some tropical fish for his aquarium. He goes to the pet store to look at some fish. The chart shows the sizes of some fish he saw.

| Species | Length |
|---|---|
| Silver Hatchet Fish | $2\frac{5}{8}$ in |
| Siamese Fighting Fish | $2\frac{1}{2}$ in |
| Black Molly | $4\frac{3}{4}$ in |
| Pearl Gourami | $3\frac{1}{4}$ in |
| Dwarf Gourami | $1\frac{7}{8}$ in |

**1.** Fred sees another fish that is $\frac{2}{3}$ the size of the Black Molly. How large is that fish?

**Step 1:** $4\frac{3}{4}$ x $\frac{2}{3}$ = ?

**Step 2:** $\frac{19}{4}$ x $\frac{2}{3} = \frac{38}{12}$

**Step 3:** $\frac{38}{12} = 3\frac{2}{12}$ inches

**Answer:** The other fish is $3\frac{2}{12}$ or $3\frac{1}{6}$ inches long.

**2.** Can we say that the Siamese Fighting fish is about $1\frac{1}{2}$ times the size of the Dwarf Gourami?

**Step 1:** $1\frac{7}{8}$ x $1\frac{1}{2}$ = ?     **Step 2:** $\frac{15}{8}$ x $\frac{3}{2} = \frac{45}{16}$     **Step 3:** $\frac{45}{16} = 2\frac{13}{16}$

**Answer:** Yes, the Siamese Fighting Fish is about $1\frac{1}{2}$ times the size of the Dwarf Gourami.

**A** Use the information in the chart to help solve these problems.

1. Other Siamese Fighting Fish may grow to be $1\frac{1}{4}$ the size of the fish Fred sees in

   the tank. How large might another fish be?

   A $3\frac{1}{8}$ in      B $2\frac{5}{8}$ in      C $3\frac{1}{2}$ in      D $2\frac{3}{4}$ in

2. Which fish is about $\frac{3}{4}$ the size of the Black Molly?

   A Dwarf Gourami          C Siamese Fighting Fish

   B Silver Hatchet Fish     D Pearl Gourami

3. What conclusion can you draw from the chart?

   A Black Mollys can grow to up to six inches long.

   B The largest fish is over twice as big as the smallest fish.

   C The first fish on the chart is about $1\frac{1}{2}$ times the size of the second.

   D The length of the second largest fish is almost four inches long.

**B** True or false? If false, write the correct answer.

1. $\frac{2}{7}$ is the reciprocal of $\frac{5}{7}$. _____

2. $2\frac{1}{2} \times \frac{1}{2} < 2\frac{1}{4} \times \frac{1}{4}$ _____

3. The reciprocal of $1\frac{3}{4}$ is $\frac{4}{7}$. _____

**C** Solve. Simplify if necessary. Show your work.

1. $\frac{1}{4} \times 3\frac{1}{3} =$          2. $1\frac{1}{2} \times 2\frac{1}{2} =$          3. $5\frac{1}{4} \times \frac{3}{4} =$

# MULTIPLYING DECIMALS

**When you multiply a whole number by a decimal, follow the same steps as when multiplying whole numbers— then place the decimal point in the product. Here's how:**

2.5
x 3
———

**Step 1:** Multiply 25 x 3 = 75

**Step 2:** Now count how many places are to the right of the decimal. In this example, there is one place to the right of the decimal.

**Step 3:** Move the decimal one place to the left, so 75 becomes 7.5

7.5 is the product of 2.5 x 3

Follow the same steps when multiplying two decimals, but count the total number of places to the right of the decimal.

0.7
x 0.9
———
0.63

 **A** Use the chart to help solve the problems.

Because Jake is concerned about the environment, he ordered recycled products from Nature's Gifts catalog.

| Item | Price |
|---|---|
| Garbage Bags | $33.50 per case |
| Napkins | $46.85 per case |
| Storage Bags | $9.25 per box |
| Tissues | $34.50 per case |
| Shopping Bag | $9.75 each |

1. When Jake orders five boxes of storage bags, he will spend—

   A $4.26          B $426.50          C $46.25          D $9.25

2. Another company sells the tissues for 1.26 times as much as Nature's Gift. How much does the other company charge for its tissues?

   A $43.47          B $34.50          C $434.70          D $4.34

3. Jake bought 1.5 cases of garbage bags. How much did he spend?

   A $33.50          B $500.25          C $5.25          D $50.25

4. If Jake buys four cases of napkins, how much will he need?

   A at least $170     B at least $180     C at least $190     D at least $220

 **Solve.**

1. 73.1 × 0.6 =          4. 18.33 × 0.06 =          7. 9.47 × 6.43 =

2. 15 × 1.5 =          5. 167.04 × 0.77 =          8. 1.99 × 2.55 =

3. 34.7 × 21.9 =          6. .09 × 0.03 =          9. 42.6 × 17.3 =

10. If beef costs $2.25 a pound, how much would 2.4 pounds cost?

11. If Ken recycles 1.3 pounds of aluminum each day, how much aluminum would Ken recycle in five days?

12. Jeanette is paid $5.67 an hour for her after-school job. How much is she paid for six hours work?

13. Maria has $30 to buy three books which cost $9.50 each. Will she have enough money?

14. Is 5.8 × 0.7 more, less, or equal to 0.58 × 7 ?

15. Bob spends $2.59 a day on snacks. How much will he spend in a week?

# MULTIPLYING DECIMALS WITH ZEROS

When multiplying two decimals, watch out for zeros in the product.

**Example:** 0.3 x 0.1 =

**Step 1:** Multiply 3 x 1 = 3

**Step 2:** Move the decimal point two places to the left.

**Step 3:** Add the zeros needed to correctly place the decimal point

$$\begin{array}{r} 0.3 \\ \times\ 0.1 \\ \hline 0.03 \end{array}$$

 Use the chart to help answer these questions.

| Date | 1999 Rainfall (Inches) |
|---|---|
| July 17 | 0.11 |
| July 18 | 0.03 |
| July 19 | 1.20 |
| July 20 | 0.56 |
| July 21 | 0.75 |

1. On July 18, 2000 it rained half as much as the previous year. Which number sentence shows how many inches it rained?

   A 0.03 x 0.5 = 0.15

   B 0.03 x 0.05 = 0.015

   C 0.03 x 1.5 = 0.015

   D 0.3 x 0.5 = 0.15

2. On July 19, the forecast had been for 0.05 as much rain as actually fell. How much rain had been forecasted?

   A 0.006 inches     B 6 inches     C 0.6 inches     D 0.06 inches

3. If it rained 1.05 times as much on July 17, 2000 than the year before, how much did it rain?

   A 1.1155 inches     B 0.1155 inches     C 11.1155 inches     D 0.01155 inches

4. On July 22, 1999, it rained 0.7 as much as the day before. How much did it rain?

    A 0.0525 inches    B 5.25 inches    C 0.525 inches    D 52.5 inches

5. Janet made the chart on page 70. She realized she recorded the incorrect amount on July 20. The actual amount was only 0.1 of the amount she wrote down. How much was the actual rainfall?

    A 0.056 inches    B 0.56 inches    C 56 inches    D 5.6 inches

 **Place the decimal point in the product. Add zeros if needed.**

1. 0.06 × 4.3 = 258

2. 1.2 × 0.07 = 84

3. 0.65 × 0.04 = 26

4. 0.03 × 5 = 15

5. 0.7 × 3.06 = 2142

6. 0.041 × 100 = 41

**C**  Solve.

1. 0.03 × .02 =

2. 13.22 × 0.5 =

3. 0.78 × 0.34 =

4. 0.9 × 0.02 =

5. 0.041 × 1000 =

6. 0.304 × 10 =

7. Carl is baking cookies for the school-wide bake sale. For every 4.2 pounds of chocolate chip cookies baked, he is baking 2.5 pounds of oatmeal cookies. If he plans to bake 21 pounds of chocolate chip cookies, how many pounds of oatmeal cookies will he bake? Make a table to solve this problem.

8. Pat worked 63 hours last month. If he made $6.75 per hour, how much did he earn?

# MIXED PRACTICE

 **A** Solve.

1. 315 x 0.24 =

3. 27.4 x 0.26 =

5. 2.94 x 1.28 =

2. 3.15 x 2.4 =

4. 2.74 x 0.126 =

6. 29.4 x 12.8 =

**B** Solve. Write the answers in simplest form.

1. $3\frac{1}{2} \times \frac{1}{2} =$

3. $\frac{1}{3} \times 6\frac{5}{9} =$

5. $7\frac{2}{7} \times \frac{11}{14} =$

2. $4\frac{5}{8} \times \frac{1}{4} =$

4. $6\frac{1}{8} \times \frac{5}{12} =$

6. $\frac{1}{3} \times 8\frac{5}{8} =$

**C** Solve.

1. 0.2 x 5 =

3. 2 x 0.025 =

5. 3.6 x 0.8 =

2. 0.4 x 0.1 =

4. 0.04 x 0.12 =

6. 0.15 x 0.09 =

**D**    Solve. Write the answers in simplest form.

1. $\dfrac{1}{7} \times \dfrac{1}{2} =$

3. $\dfrac{2}{9} \times \dfrac{3}{5} =$

5. $\dfrac{3}{8} \times \dfrac{1}{2} =$

2. $\dfrac{3}{4} \times \dfrac{5}{7} =$

4. $\dfrac{2}{5} \times \dfrac{2}{5} =$

6. $\dfrac{5}{6} \times \dfrac{1}{3} =$


**E**    Solve. Write the answers as mixed numbers in simplest form.

1. $2\dfrac{1}{3} \times 9\dfrac{1}{2} =$

3. $4\dfrac{1}{5} \times 5\dfrac{1}{2} =$

5. $8\dfrac{4}{9} \times 1\dfrac{5}{6} =$

2. $3\dfrac{2}{5} \times 4\dfrac{4}{9} =$

4. $2\dfrac{6}{7} \times 4\dfrac{1}{4} =$

6. $1\dfrac{3}{4} \times 8\dfrac{3}{4} =$


**F**    Change each mixed number to an improper fraction.

1. $6\dfrac{1}{8} =$

3. $1\dfrac{3}{5} =$

5. $8\dfrac{3}{4} =$

2. $7\dfrac{1}{7} =$

4. $9\dfrac{2}{3} =$

6. $4\dfrac{5}{6} =$

## TEST TRACK

**There are many types of screws that come in all lengths. Here are some examples:**

**A**  Choose the correct answer for the questions below using this chart.

| | Type of Screw | Minimum Size | Maximum Size |
|---|---|---|---|
| | **Machine Screw** | $\frac{5}{8}$ inch | $1\frac{3}{4}$ inches |
| | **Wood Screw** | $\frac{1}{2}$ inch | 3 inches |
| | **Set Screw** | 1 inch | 3 inch |
| | **Cap Screw** | $\frac{1}{2}$ inch | 3 inches |

1. If a new machine screw is $1\frac{1}{4}$ times larger than the current maximum size, what will its length be?

   Ⓐ $2\frac{1}{16}$ inches      Ⓑ $2\frac{3}{8}$ inches      Ⓒ $1\frac{4}{8}$ inches      Ⓓ $2\frac{3}{16}$ inches

2. Chris wants a cap screw that is 0.75 of the maximum size. How long will that cap screw be?

   Ⓐ 1.50 inches      Ⓑ 2.25 inches      Ⓒ 2.50 inches      Ⓓ 3.00 inches

3. For wood screws, there is a length $1\frac{1}{8}$ larger than the minimum. How long is that length?

   Ⓐ $\frac{9}{16}$ inches      Ⓑ $\frac{11}{16}$ inches      Ⓒ 1 inch      Ⓓ $1\frac{1}{2}$ inches

4. Lin needs a wood screw that is twice the length of the minimum size. How long will that wood screw be?

   Ⓐ $\frac{1}{2}$ inch      Ⓑ $\frac{1}{3}$ inch      Ⓒ 2 inches      Ⓓ 1 inch

**B** Multiply. Simplify if necessary.

1. $0.07 \times 0.6 =$

2. $14.3 \times 0.02 =$

3. $1\frac{2}{3} \times 3\frac{4}{5} =$

4. $2\frac{1}{3} \times 4\frac{1}{6} =$

5. $\frac{3}{4} \times 1\frac{3}{4} =$

6. $\frac{3}{8} \times 2\frac{1}{3} =$

7. $\frac{3}{5} \times 1\frac{1}{2} =$

8. $3\frac{1}{5} \times \frac{2}{3} =$

9. $0.03 \times 1.9 =$

10. $24.7 \times 3.3 =$

11. $0.04 \times 0.002 =$

12. $1000 \times 0.003 =$

**C** True or false? If false, give the correct answer.

1. $2.38 \times 0.05 = 0.119$ _____

2. In the multiplication problem $\frac{3}{4} \times \frac{5}{9} = \frac{15}{36}$, $\frac{3}{4}$ is a product. _____

3. Does $\frac{6}{7} \times \frac{5}{7} = \frac{30}{49}$ ? _____

4. $1\frac{4}{9} \times 1\frac{2}{3} = 3\frac{5}{9}$ _____

# DIVIDING FRACTIONS

Jane needs 4 cups of milk for a recipe she is trying. But she only has a $\frac{1}{2}$ cup measuring cup. How many half-cups are in 4 cups?

Question: How many half-cups are in 4 cups?

Solve: $4 \div \frac{1}{2} =$

To divide by a fraction, multiply by the reciprocal of the divisor.

$4 \div \frac{1}{2} = \frac{4}{1} \times \frac{2}{1} = 8$   Jane can fill 8 half-gallon jugs.

Next year, Paul hopes to grow more Chrysanthemums. The number he has growing now in his garden is $\frac{1}{5}$ of what he hopes to grow. How many chrysanthemums does Paul plan to grow?

| Flower | Quantity |
|---|---|
| Chrysanthemum | 20 |
| Poppy | 35 |
| Lily | 10 |
| Yarrow | 5 |
| Bellflower | 18 |

Question: How many chrysanthemums does Paul plan to grow?

Solve: $\frac{20}{1} \div \frac{1}{5} =$

$\frac{20}{1} \times \frac{5}{1} = 100$

Paul hopes to grow 100 chrysanthemums.

**A**   Use the chart to help solve these problems.

1. Paul loves to give away flowers, especially lilies. The number of lilies remaining in the garden is listed in the chart above. It is just $\frac{1}{10}$ of the number he had last year. How many lilies were in Paul's garden last year?

   A  10          B  100          C  1          D  1000

2. Paul can make three bunches of bellflowers. If he plans to give each person $\frac{1}{4}$ of a bunch, how many people can he give bellflowers to?

   A  18          B  3          C  12          D  4

3. Last year, Paul's flowers were more plentiful. This year, his poppies are just $\frac{1}{2}$ of the amount he had last year. How many poppies did Paul have last year?

   A  75          B  35          C  50          D  70

4. The number of chrysanthemums in Paul's garden is $\frac{2}{5}$ as many as his friend Cindy is growing. How many chrysanthemums does Cindy have?

   A  500          B  100          C  5          D  50

**B**  Complete the number sentences.

1.  $15 \div \frac{1}{3} =$

2.  $4 \div \frac{1}{4} =$

3.  $\frac{1}{8} \div \frac{3}{10} =$

4.  $5 \div \frac{1}{2} =$

5.  $\frac{2}{7} \div \frac{1}{7} =$

6.  $7 \div \frac{1}{5} =$

7.  $4 \div \frac{1}{2} =$

8.  $18 \div \frac{1}{6} =$

9.  $\frac{1}{3} \div \frac{1}{12} =$

10. $\frac{2}{3} \div \frac{1}{2} =$

**C**  Solve. Show your work.

1. Bonnie makes 20 cups of her special snack mix for her birthday party. If she plans to give each person $\frac{1}{3}$ cup of the mix, how many people can she feed?

2. At his barbecue, Kyle is planning to make $\frac{1}{4}$ pound hamburgers. If he has six pounds of meat, how many hamburgers can Kyle make?

# DIVIDING MIXED NUMBERS

Dividing by a mixed number is very similar to dividing by a fraction.

**Example:** $5 \div 1\frac{1}{4} =$

**Step 1:** Convert the mixed number to an improper fraction: $1\frac{1}{4} = \frac{5}{4}$

**Step 2:** Write the reciprocal of the divisor and multiply.

$$\frac{5}{1} \times \frac{4}{5} = \frac{20}{5} = 4$$

**Answer:** $5 \div 1\frac{1}{4} = 4$

| The Appalachian Trail | |
|---|---|
| **State** | Miles of Trail |
| **Tennessee** | $88\frac{7}{10}$ |
| **Georgia** | $75\frac{2}{5}$ |
| **New Hampshire** | $117\frac{1}{10}$ |
| **Pennsylvania** | $229\frac{2}{5}$ |
| **New Jersey** | $73\frac{2}{5}$ |
| **New York** | $88\frac{4}{5}$ |

 **Use the chart to answer the following questions.**

The Appalachian Trail is a hiking trail covering 2,158 miles and crossing 14 states. Most people take five to six months to hike the entire trail. The chart shows some of the states it crosses.

1. Joey hiked 14 miles per day.
   How long did it take him to hike through Tennessee?

   A more than 6 days     C less than 6 days

   B more than 8 days     D less than 5 days

2. If it took a hiker four days to hike through Georgia, how far did he hike each day?

   A more than 20 miles     C more than 25 miles

   B less than 18 miles     D more than 18 miles

3. If a hiker takes 5 days to cross Tennessee, how many miles does she hike each day?

   A  less than 17

   B  more than 17

   C  less than 15

   D  more than 20

4. If a hiker wants to cross New Jersey in 4 days, how far must he hike each day?

   A  less than 16 miles

   B  less than 18 miles

   C  more than 19 miles

   D  more than 18 miles

5. If Stu walked $16\frac{1}{2}$ miles each day, how long did it take to cross Georgia?

   A  more than 7 days

   B  less than 4 days

   C  more than 4 days

   D  less than 2 days

6. Joy hiked 20 miles per day through New Hampshire. How long did it take her to cross New Hampshire?

   A  almost 4 days

   B  almost 5 days

   C  almost 6 days

   D  almost 7 days

**B**  Solve.

1. $4\frac{1}{2} \div 1\frac{1}{2} =$

2. $6 \div 2\frac{1}{6} =$

3. $5\frac{1}{8} \div \frac{2}{3} =$

4. $8\frac{1}{2} \div 2 =$

5. $3\frac{1}{5} \div 1\frac{2}{3} =$

6. $6\frac{1}{4} \div \frac{2}{3} =$

7. $3\frac{2}{3} \div 2\frac{1}{5} =$

8. $5 \div 1\frac{1}{4} =$

9. $7\frac{3}{4} \div 2 =$

**C**  Write a number sentence and solve each of the following problems.

1. Carly is cutting a six-foot board into $1\frac{1}{2}$ foot pieces. How many pieces can she cut?

2. A piece of cloth is $3\frac{3}{4}$ yards in length. If you wanted to cut the cloth into $1\frac{1}{4}$ yard sections, how many sections could you cut?

Dividing Mixed Numbers    79

# DIVIDING A DECIMAL BY A WHOLE NUMBER

**To divide a decimal by a whole number follow these steps.**

**Example:** $247.5 \div 5 =$

**Step 1:** Set up the problem as you would when dividing any numbers: $5\overline{)247.5}$

**Step 2:** Bring up the decimal point and divide:

$$
\begin{array}{r}
49.5 \\
5\overline{)247.5} \\
\underline{2\,0}\phantom{.5} \\
47 \\
\underline{45} \\
25 \\
\underline{25}
\end{array}
$$

| Event | Winning Weight in Pounds |
|---|---|
| **Bench Press** | 363.7 |
| **Deadlift** | 639.2 |
| **Squat** | 534.4 |
| **Curl** | 137.3 |

**A** Use the chart to solve these problems.

Mike and his family went to the weightlifting competition at the Coliseum.

1. If the second place finisher in the curl lifted a total of 270.6 pounds for two curls, how much did the weightlifter lift each time?

   A 137.3 pounds    B 135.3 pounds    C 13.53 pounds    D 1353 pounds

2. The leading competitor in the deadlift lifted a total of 1,910.4 pounds in three lifts. How much was each lift?

   A 636.8 pounds    B 63.68 pounds    C 6,268 pounds    D 6.368 pounds

3. How much total weight would the winner in the squat competition lift if all four tries were equal?

   A 2,137.6 pounds    B 213.76 pounds    C 21,376 pounds    D 2,317.6 pounds

4. A beginner in the squat needed two tries to equal the winner. How much did the beginner lift each time?

   A 267.2 pounds    B 26,72 pounds    C 2,672 pounds    D 2.672 pounds

5. The Bench Press winner lifted a total of 722.4 pounds on his next two tries. How much did he lift each time?

   A 361.2 pounds      B 36.12 pounds      C 3,612 pounds      D 3.612 pounds

**B** True or false? If false, provide the correct answer.

1. $16.5 \div 3 = 5.5$

2. $54.6 \div 7 = 78$

3. $360.4 \div 21.2 = 17$

4. $109.8 \div 6 = 1.83$

5. $12.009 \div 3 = 4.3$

6. $65.7 \div 9 = 7.3$

**C** Solve. Show your work.

1. 37 tickets for the school play were sold for a total of $194.25. How much did each ticket cost?

2. Packages of paper are on sale for 5 packages for $8.75 or 10 packages for $15.00. Which is the better deal? Explain your answer.

3. Stan made 19.2 pounds of fudge. If he gave the same amount of fudge to 12 friends, how much did each person get?

4. Create and solve your own word problem using the number sentence $25.2 \div 4 =$ .

## DIVIDING BY A DECIMAL

**Here's how to divide a decimal by another decimal.**

**Example:** 38.18 ÷ 4.6 =

**Step 1:** Convert the divisor (4.6) into a whole number (46) by moving the decimal one place to the right.  $4.6\overline{)\phantom{0000}}$

**Step 2:** How many decimal places are in the original divisor? (1)

**Step 3:** Move the decimal point in the dividend to the right by the same number of places. Bring the decimal point up and divide:

$$\begin{array}{r} 8.3 \\ 4.6\overline{)38.18} \\ \underline{368} \\ 138 \end{array}$$

| Country | CD's per person |
|---------|-----------------|
| USA | 3 |
| Japan | 2.3 |
| France | 1.8 |
| Brazil | 0.5 |
| Canada | 1.7 |
| Spain | 0.9 |

*Source: http://www.mebis.com*

**A** Use the chart to answer these questions.

1. Sales of CDs in France reached 101,574,000 last year. How many people live in France?

   A 101,574,000

   B 10,157,400

   C 56,430,000

   D 15,574,000

2. The number of people in Australia buying CDs is 35.3 million. If 70.6 million CDs are sold, how many CDs does the average person buy?

   A  1          B  2          C  3          D  4

3. In Japan, 287.5 million CD's are bought every year. How many people live in Japan?

   A  25 million      B  125 million      C  2.5 million      D  12.5 million

4. Canadians buy 49.3 million CD's. About how many people live in Canada?

   A  1.9 million      B  2.9 million      C  19 million      D  29 million

**B**  Solve. Show your work.

1. $5.4 \div 4 =$          4. $3.4 \div 4$          7. $0.52 \div .8$

2. $5.35 \div 2 =$          5. $0.77 \div .5$          8. $3.9 \div 1.5$

3. $0.9 \div 5$          6. $10.75 \div 2$          9. $16.73 \div 1.4$

10. If Sam slices 0.5 pounds of cheese into 4 equal pieces, how much does each piece weigh?

11. A playground 39.2 yards wide is being divided into sections 5.6 yards wide for field day. How many sections will there be?

12. A piece of cloth 25.2 feet long is being cut into pieces 1.4 feet long for an art project. How many pieces will there be?

13. Mary had 1.8 yards of ribbon to cut into five pieces for gift wrapping. How long was each piece?

**TEST TRACK**

Many towns in southwestern Colorado are known for skiing. Here's how much snow they get in an average year:

| Town | Elevation | Snowfall |
|------|-----------|----------|
| Cortez | 5,911 feet | 39.2 inches |
| Durango | 6,685 feet | 65.4 inches |
| Ouray | 7,730 feet | 142.3 inches |
| Silverton | 9,430 feet | 154.3 inches |

**A** Chose the answer to each question.

1. If the snowfall in Cortez occurs over eight months, what is the average snowfall per month?

   Ⓐ 0.49 inches    Ⓑ 49 inches    Ⓒ 4.9 inches    Ⓓ 3.92 inches

2. If $26\frac{1}{2}$ inches of snow falls in 1 week in Silverton, what is the average daily snowfall for the week?

   Ⓐ $2\frac{5}{7}$ inches    Ⓑ $3\frac{11}{14}$ inches    Ⓒ $3\frac{5}{7}$ inches    Ⓓ $3\frac{6}{7}$ inches

3. If .5 inches of snow fell in Durango in two hours, how many inches fell per hour?

   Ⓐ 2.5    Ⓑ 0.25    Ⓒ 0.025    Ⓓ 25.0

4. What conclusion can you draw from the chart?

   Ⓐ Lower elevation means more snow.

   Ⓑ Small towns get more snow.

   Ⓒ The higher the elevation, the greater the snowfall.

   Ⓓ Other towns get more snow.

**B** Complete the following number sentences.

1. $\dfrac{4}{5} \div \dfrac{1}{3} =$

2. $2\dfrac{4}{5} \div 1\dfrac{1}{2}$ ◯ $3\dfrac{1}{2} \div \dfrac{1}{2}$

3. $5.46 \div 6 =$

4. $36.45 \div 5$ ◯ $29.2 \div 4$

5. $28.42 \div 5.8 =$

6. $5.50 \div 2$ ◯ $6.45 \div 3$

7. $\dfrac{2}{3} \div \dfrac{2}{3}$ ◯ $\dfrac{3}{4} \div \dfrac{3}{4}$

8. $6\dfrac{3}{8} \div 2\dfrac{5}{6} =$

9. $43.29 \div 5$ ◯ $4.329 \div .5$

10. $97.17 \div 23.7 =$

**C** Solve.

1. Byron's driveway was 33.2 feet long. He wanted to divide it into four sections to refinish it. How long would each section be?

2. Jen had 6 feet of ribbon to wrap gifts. If she wanted to wrap 12 gifts, how much ribbon could she use for each gift?

3. Pat has 1 4/5 gallons of juice which he divided into $\dfrac{1}{5}$ gallon containers. How many containers can he fill?

4. Baseball fans spent $197 for programs. If the programs cost $0.50 each, how many programs were sold?

5. Write a word problem based on this number sentence: $3\dfrac{3}{4} \div \dfrac{1}{2} =$ .

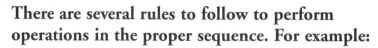

# ORDER OF OPERATIONS

**There are several rules to follow to perform operations in the proper sequence. For example:**

$32 - (10 \div 5) \times 9 \div 3 =$

Step 1: Perform the operation inside the parenthesis first. $(10 \div 5) = 2$

Step 2: Perform any multiplication or division from left to right: $2 \times 9 = 18 \div 3 = 6$

Step 3: Perform any addition or subtraction from left to right: $32 - 6 = 26$

Mike bought some gift baskets at the following prices:

| Description | Price |
|---|---|
| Gourmet Feast | $59.75 |
| Cookie Delight | $64.50 |
| Candy Basket | $49.25 |
| Fruits of the World | $74.00 |

Mike ordered two Candy Baskets and A Cookie Delight. How much change will he receive if he pays with four $50 bills?

Step 1: $49.25 \times 2 + $64.50 = $163 cost of items

Step 2: $4 \times 50 = $200 amount paid

Step 3: $200 - $163 = $37 change received

**A** Use the chart above to answer these questions.

1. Mike ordered two Gourmet Feasts and a Candy Basket. He split the cost with two friends. Which number sentence shows how much Mike and his friends each spent?

A ($59.75 \times 2 + 49.25) \div 3 =$     C $59.75 \times (2 + 49.25 \times 3) =$

B ($59.75 \times 2) + (49.25 \div 3) =$     D $59.75 \times 2 + 49.25 \times 3 =$

2. What is the answer to question 1?

   A $56.25    B $5.65    C $359.70    D $65.55

3. At next week's half-price sale, Mike plans to buy two Gourmet Feasts and two Cookie Delights. How much will he save off the regular price?

   A $124.25    B $59.75    C $135.25    D $210.00

4. If Mike spends a total of $203 at the regular price, what did he buy?

   A  two Cookie Delights and one Fruits of the World

   B  two Candy Baskets and two Cookie Delights

   C  two Fruits of the World and One Cookie Delight

   D  two Gourmet Feasts and one Fruits of the World

## B  Solve these equations.

1. $16 + (12 - 3) =$

2. $30 - (10 \times 2) =$

3. $15 + (23 \times 7) - 12 =$

4. $(48 \div 4) \times 2 + 18$

True or false?  If false, please correct.

5. $6 \times (40 + 50) = 96$

6. $99 + (15 \times 3 \div 5) = 108$

7. $62 + (7 \times 7) - 9 = 474$

8. $92 - (98 \div 7 \times 4) = 36$

## C  Solve these word problems.  Show your work.

1. John bought four bags of groceries at $43.45 each.  He also picked up two magazines at $1.95 apiece.  If he gave the cashier $200, how much change did he receive?

2. At the end of the day, Jack had 600 balloons left.  He had sold $\frac{1}{2}$ of his original supply at the circus and $\frac{1}{3}$ of his original supply at a street fair.  How many balloons did Jack start out with?

# CALCULATOR REVIEW

**A**  Use your calculator to find the sums. Arrange the letters of the number sentences from least to greatest.

1. A  63.2 + 15.01 + 41.003  =

   B  61.02 + 15.3 + 40.67  =

   C  2.98 + 15.006 + 41.11  =

2. A  85 + 0.7 + 0.004 + 6.9  =

   B  0.82 + 0.5 + 0.009 + 7.1  =

   C  0.86 + 0.6 + 0.008 + 7.0  =

_____

**B**  Use the chart below and your calculator to help solve these problems.

Every year, The Food Marketing Institute tracks average supermarket sales per week as shown in this chart.

| Year | Average Weekly Sales Per Supermarket |
|------|--------------------------------------|
| 1992 | $183,200 |
| 1993 | $192,760 |
| 1994 | $193,035 |
| 1995 | $209,875 |
| 1996 | $212,382 |
| 1997 | $284,700 |
| 1998 | $333,411 |
| 1999 | $334,479 |

*Source: Food Marketing Institute*

1. What were the total sales for the last four years shown on the chart?

   A  $778,870

   B  $1,164,972

   C  $1,346,893

   D  $875,972

2. How much greater were average sales for the last four years when compared to the first four years shown on the chart?

   A  $1,943,842     B  $778,870     C  $1,164,972     D  $386,102

**C**  Solve. Use your calculator.

1. 76,456 + (96 × 154) =

2. (973,546 - 1,269) + 3,786 =

3. 5,455 + 6,777 + (420 ÷ 28) =

4. (45,119 - 19) × (347 - 285) =

5. 23 + 45 + (36 × 8) ÷ 2) =

**D**  Enter the correct sign: < > =.

6. (456,376 - 53,987) + 322,145 ◯ (390,511 + 28,362) - 64,988

7. 2,677 + 45,432 + (1,723 - 574) ◯ 53,000 - 2,367

8. (26,678 - 3,999) + (56,456 - 765) ◯ (32,786 - 9,123) + (63,447 - 12,781)

9. 47,000 - (15,000 + 8,000) + 40,000 ◯ (6,000 - 2,000) + (70,000 -15,000)

10. (872 + 147) - (543 + 674) + 22 ◯ (1,000 + 535 - 89) - (72 + 333)

  **E**  Answer the questions using the chart below.

Every July, a three-week bicycle race takes place in France, known as the Tour de France. The race attracts top riders from all over the world. Here were the standings after two weeks of the race in July 2000.

1. What is the difference in the combined times between the top three riders and the next three riders?

2. Is the combined time of the French riders more or less than the combined time of the German and Austrian riders?

| Rider | Country | Time (Hours) |
|---|---|---|
| Armstrong | U.S.A. | 49 |
| Ulrich | Germany | 54 |
| Beloki | Spain | 55 |
| Moreau | France | 56 |
| Beltran | Spain | $56\frac{1}{2}$ |
| Virenque | France | 57 |
| Heras | Spain | $57\frac{1}{2}$ |
| Mancebo | Spain | 59 |
| Ochoa | Spain | $59\frac{1}{2}$ |
| Luttenberger | Austria | $59\frac{1}{2}$ |

# SOLVING WORD PROBLEMS

**LESSON 3 UNIT 8**

**Games of July 22**

| Teams | Fans |
|---|---|
| Reds - Diamondbacks | 42,568 |
| Cardinals - Astros | 42,588 |
| Tigers - Royals | 35,046 |
| Indians - Twins | 14,168 |

**A**    Use the chart above to answer these questions.

1. The fans at the Reds - Diamondbacks game order an average of two drinks each. The attendance at the next game is 41,791 and the fans average the same number of drinks. To find the difference in the total number of drinks sold for each game, we would use:

   A  42,568 - 41,791 x 2 =      C  42,568 x 2 - 41,791 x 2 =

   B  42,568 + 41,791 x 2 =      D  2 x 42,568 x 41,791 =

2. There are six seating sections in the upper deck of Royals stadium. The first section has 2,400 seats; the second section, 2,700; and the third section, 3,100. If the pattern continues, what is the seating capacity of the upper deck of the Royals' ballpark?

   A  20,600     B  18,800     C  14,400     D  20,900

3. Which question cannot be answered using the chart?

   A  What was the average attendance for all four games?

   B  How many fans attended all four games?

   C  How many seats were empty at the Reds - Diamondbacks game?

   D  What was the difference in attendance at the Tigers- Royals game and the Cardinals - Astros game?

4. There are 1,125 fans in Section 5 at the Indians - Twins game. On average, every adult has two children with them. Section 6 is three times as large as section 5. If the ratio of adults to children is the same, how many children are sitting in both sections?

    A  3,000       B  2,250       C  3,375       D  2,500

**B** Answer true or false to the following statements. If false, write the correct answer.

1. Karen thought she could climb Kilimanjaro in eight days and Mt. Elbrus in four days. She thinks she will need two gallons of water per day. To calculate how much water she'll need to climb both mountains, Karen can use this number sentence: $(8 + 2) \times 4 = 64$ gallons.

2. A climber on Mt. Denali is traveling with a group of 25 experienced mountain climbers. There are seven more men than women in the group. There are 16 men in the group.

3. One climber spent $3,395 to climb Aconcagua. If he spent $875 for camping equipment, $1,035 for travel, and $560 for food, he paid $925 for other supplies.

4. Puncak Jaya (19,750 ft high) can be climbed at a rate of about 4,200 feet per day if you climb $8\frac{1}{2}$ hours per day. There will be about 40 hours of climbing involved.

5. Most climbers on Mt. Kilimanjaro (19,340 ft high) can climb an average of 2,147 feet per day, but they need help carrying supplies. The helpers charge $25 per day. If there are six helpers, the cost of the helpers could be found using this number sentence: $(19{,}340 \div 2{,}147) \times 6 \times 25 =$

6. To prepare for a climb up the Vinson Massif a hiker started preparing six months in advance. The first month, he walked 240 miles wearing a heavy backpack. By the fifth month, he had totaled 1,400 miles. If he followed a consistent pattern, he increased his mileage by 30 miles each month.

# SOLVING WORD PROBLEMS

**A recent study was conducted on the highest level of education achieved by American Adults.**

| Category | Percentage of Americans over Age 25 |
|---|---|
| 4 years or less of Elementary School | 2.00% |
| 5 - 8 years of Elementary School | 6.00% |
| 1 - 3 years of high school | 10.00% |
| 4 years of high school | 32.50% |
| 1 - 3 years of college | 24.00% |
| 4 years or more of college | 20.25% |

*Source: U.S. Census Bureau, Economic Policy Institute*

 **A** Use the chart above to answer these questions.

1. If the percentage of Americans completing college today is $2\frac{1}{4}$ times what it was 30 years ago, what was the percentage 30 years ago?

   A 9%          B 19%          C 10%          D 13%

2. The percentage of Americans completing 5 - 8 years of elementary school was 3 times higher in 1970. In 1980, it was $\frac{5}{9}$ times the percentage in 1970. What was the percentage in 1990 if it was $\frac{4}{5}$ the percentage in 1980?

   A 8%          B 9%          C 7%          D 6%

3. In Culver County, 6,428 adults completed one to three years of high school in one town. A neighboring county had 1.5 times more adults completing one to three years of high school. If the high schools in both counties average 1,100 students, how many high schools are there?

   A about 15    B about 10    C about 20    D about 25

4. Which question can not be answered using the information on the chart?

   A  What is the highest education level of most American adults?

   B  How many adults did not complete college?

   C  How many adults completed 5 to 8 years of elementary school?

   D  What increases in education levels have occurred the last 30 years?

**B**   **Solve the following problems.**

1. In two days, it snowed $16\frac{1}{4}$ inches. It snowed $1\frac{1}{4}$ inches more the second day than the first. How much did it snow each day?

2. Hank bicycled $10\frac{1}{2}$ miles on Monday and $11\frac{7}{10}$ miles on Tuesday. If he bicycled a total of $34\frac{1}{2}$ miles in three days, how far did Hank bicycle on Wednesday?

3. In the after-school sports program, 20 children chose to play baseball. Half that many chose golf, and three times as many children signed up for basketball as compared to golf. The number of children choosing baseball was half the number of children choosing volleyball. What percent of the students chose volleyball?

4. Anessa sold 300 chocolate chip cookies at the bake sale for $420. She spent $140 for flour, eggs, sugar, and chocolate. If she buys $\frac{1}{4}$ as many ingredients as the first time, how many cookies can she make for the next bake sale?

5. Gene had $45.25 last week. He worked four days and now has $165.25. If he worked four hours each day, how much did he earn per hour?

6. There are 42 students in the school choir. If there are $\frac{3}{4}$ as many boys as girls, how many boys and girls are in the school choir?

7. Terri has eight dresses which sell for $24.95 each. If she sold $374.25 worth of dresses yesterday, how many dresses did she sell? How many dresses did she have before yesterday?

8. Laura spent $6.95 for a plant and $5.25 less than that for a package of seeds. If she spent $23.40 in all, how much did Laura spend on gardening tools?

The other day there was a big truck race. Here are some of the results:

**Top Five Finishers in Michigan 200**

| Driver | Winnings |
|---------|----------|
| Biffle | $52,090 |
| Busch | $35,555 |
| Wallace | $26,950 |
| Houston | $21,570 |
| Setzer | $19,050 |

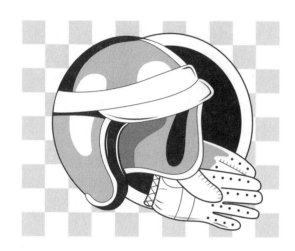

 Use the chart above to answer the following questions.

1. The combined winnings of the sixth and seventh place finishers were 72% of the combined winnings of the fourth and fifth place finishers. If the sixth place finisher made $850 more than the seventh place finisher, how much did the seventh place finisher win?

   Ⓐ about $14,200        Ⓒ about $14,500

   Ⓑ about $13,700        Ⓓ about $13,900

2. Which question cannot be answered using this chart?

   Ⓐ How much did the top five finishers win?

   Ⓑ How much did the winner earn in his last race?

   Ⓒ What was the difference in winnings between second and third place?

   Ⓓ How much less did the fifth place finisher win than the first place finisher?

3. Of all the racers, there are $1\frac{1}{2}$ times more veterans than rookies. If there are 45 racers, how many veterans and rookies were there?

   Ⓐ 20 rookies, 25 veterans          Ⓒ 21 rookies, 24 veterans

   Ⓑ 16 rookies, 29 veterans          Ⓓ 18 rookies, 27 veterans

4. In his next race, Biffle won 10% more money. If Biffle and Busch won $99,364 together, how much did Busch win?

   Ⓐ $42,065          Ⓒ $39,547

   Ⓑ $46,024          Ⓓ $43,820

## B    Solve.

1. 348 students paid a total of $82,302 for lessons at the Jones Art Institute. At the Myers Art School, 285 students paid $70,466.25 for lessons. Which school offered a better price? What was the difference per student?

2. Using the number sentence $(3\frac{2}{3} \times 1\frac{1}{8}) + 5\frac{1}{4}$, write and solve your own word problem.

3. Three buses are traveling on a school trip. The last bus is carrying 32 passengers. Each bus ahead of it is carrying four less than $1\frac{1}{2}$ times the bus behind of it. How many students are in the three buses?

4. Vince bought four sports shirts for $12.95 each. He got a discount of 20% for buying more than three. If Vince has two $20 bills, how much more money will he need?

5. The first week, a plant grew by $\frac{1}{2}$ an inch, the second week, $\frac{2}{3}$ of an inch; and the third week, $\frac{5}{6}$ of an inch. If the pattern continues, how tall will the plant be by the fifth week?

6. Ted is trying to lose weight on a new diet. The first week he lost 1.2 pounds. The second week, he gained 0.5 pounds. The third week, he lost 1.5 pounds, and the fourth week he gained 0.6 pounds. If the pattern continues, how much weight will he lose in eight weeks?

# GLOSSARY

## A

**addend** - A number that is added to another number to find a sum is called an addend. In 4 + 6 = 10, 4 and 6 are addends; 10 is the sum.

## D

**decimal number** - A decimal is a fraction with a denominator of 10 or a power of 10 expressed using a decimal point instead of a denominator. For example, 7/10 = 0.7, 34/100 = 0.34, and 3/1000 = 0.003

**denominator** - The number that is below or to the right of the line in a fraction is called the denominator. In 2/7, 7 is the denominator.

**digit** - any of the counting numbers from 0 to 9

**difference** - The amount that is left after one number is subtracted from another is called the difference. In 11-6 = 5, 5 is the difference.

**dividend** - A number that is divided by another number is called the dividend. In 21 ÷ 7 = 3, 21 is the dividend.

**divisor** - A number by which another number is divided is called the divisor. In 36 ÷ 9 = 4, 9 is the divisor.

## F

**factor** - A number that is multiplied by another number or numbers to find a product is called a factor. In 3 x 4 x 6 = 72, 3, 4, and 6 are factors.

## L

**like fractions** - Fractions which have the same denominator are said to be like fractions. 1/6 and 5/6 are like fractions.

## M

**minuend** - The number from which another number is subtracted is called the minuend. In 7 - 3 = 4, 7 is the minuend.

## N

**numerator** - The number that is above or to the left of the line in a fraction is called the numerator. In 2/7, 2 is the numerator.

## P

**percent** - A hundredth of a number is a percent. The symbol for percent is %. 8/100 is 8%.

**product** - The amount that results from multiplying two or more numbers together is called the product. In 2 x 3 = 6, 6 is the product. 2 and 3 are factors.

## Q

**quotient** - The amount that results from dividing one number by another is called the quotient. In 14 ÷ 7 = 2, 2 is the quotient. 14 is the dividend and 7 is the divisor.

## S

**subtrahend** - A number subtracted from another number is called the subtrahend. In 6 - 4 = 2, 4 is the subtrahend. 6 is the minuend and 2 is the difference.

**sum** - The result of adding two or more addends is called their sum. In 2 + 4 + 5 = 11, 11 is the sum. 2, 4, and 5 are addends.

## U

**unlike fractions** - Fractions which have different denominators are said to be unlike fractions. 1/7 and 5/6 are unlike fractions.

## W

**whole number** - Any number zero and greater containing no fractions or decimals is called a whole number. 0, 1, 2, 3, 4 . . . are whole numbers.